DATE DUE

LET'S INVESTIGATE SCIENCE

Electricity and Magnetism

LET'S INVESTIGATE SCIENCE

Electricity and Magnetism

Robin Kerrod
Illustrated by Ted Evans

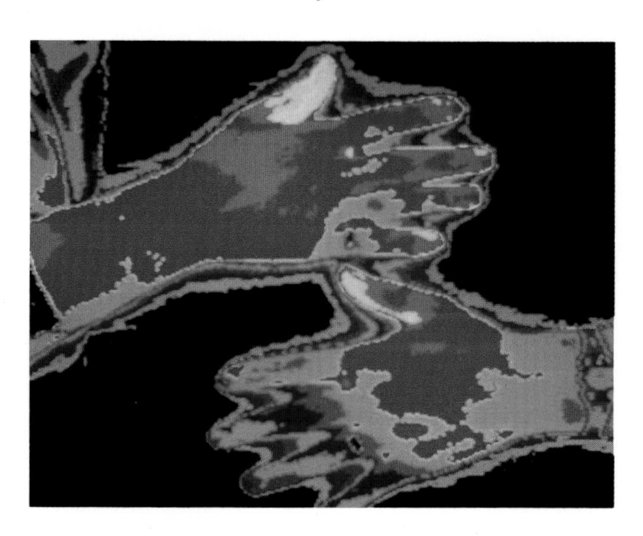

MARSHALL CAVENDISH
NEW YORK · LONDON · TORONTO · SYDNEY

Library Edition Published 1994

Published by Marshall Cavendish Corporation
2415 Jerusalem Avenue
PO Box 587
North Bellmore
New York 11710

Series created by Graham Beehag Book Design

Library of Congress Cataloging-in-Publication Data

Kerrod, Robin.
Electricity and magnetism/ Robin Kerrod; llustrated by
Ted Evans.
p. cm. -- (Let's investigate science)
Includes index.
ISBN 1-85435-626-7 ISBN 1-85435-688-7 (set)
1. Electricity--Juvenile literature. 2. Electricity--Experiments--Juvenile literature. 3. Magnetism--Juvenile literature. 4. Magnetism--Experiments--Juvenile Literature. [1. Electricity. 2. Electricity--Experiments. 3. Magnetism. 4. Magnetism--Experiments. 5. Experiments.]
I. Evans, Ted ill. II. Title. III. Series: Kerrod, Robin. Let's investigate science.
QC527.2..K47 1994 93-46013
537--dc20 CIP
AC

Printed and bound in Hong Kong

Contents

Introduction

Electricity literally holds the Universe together, because all matter is essentially electrical in nature. But in the natural world we see this universal electricity only occasionally.

Most notably we see it in the brilliant streaks of lightning that play among the clouds and stab at the ground during thunderstorms. Lightning packs tremendous energy, which can fell trees, demolish buildings, and kill animals and people.

Fortunately, we do not have to rely on nature to provide us with electricity, which has become our most useful and most convenient form of energy. We can make it ourselves, using batteries and generators. Batteries use chemistry to produce electricity. Generators work by taking advantage of the close relationship between electricity and magnetism. Magnetism is a natural property of iron and a few other metals.

In this book we investigate the nature of electricity and magnetism. We see how they are used in our everyday lives, in science, and in industry, to create devices from light bulbs and microchips to furnaces and atom-smashers.

You can check your answers to the questions featured throughout this book on pages 60-61.

◀ Robot arms carry out electric-arc welding on a production line. They are guided by the trickle of electrons through the miniature electric circuits of their microchip controllers. Such electronic robots weld with greater precision than human workers do, and they never tire or complain of heat, fumes, or noise.

1 The Big Attraction

This chapter is concerned with a kind of electricity we call static electricity, and with magnetism. Static electricity has to do with electrical charges that attract and repel one another. Magnetism has to do with magnetic poles that attract and repel one another.

Static electricity contrasts with the more familiar form of electricity, known as current electricity. This is the kind that flows in wires (see Chapter 2).

The word electricity comes from "elektron," the Greek word for amber. Amber is a substance that acquires a static electric charge when it is rubbed. It was used by scientists in early electrical experiments. The word magnetism is named for a place in Asia called Magnesia, where stones were found that were natural magnets.

These magnetic stones, called lodestones ("guiding stones"), came into use as compasses for navigation some 900 years ago. Compasses work because the Earth itself behaves as if it contained a huge magnet.

◄◄ Man-made lightning leaps between switches during an experiment being carried out at Sandia National Laboratories near Albuquerque, New Mexico. The power builds up in this apparatus to 30 terawatts.

Q 1. "Tera" means million million. So how many ordinary 100 watt light bulbs could be run from the apparatus?

◄ U.S. scientist Benjamin Franklin risked his life when he flew a kite in a thunderstorm in 1752 to prove that lightning was electricity. Lightning traveled down the kite wire to a key and made a spark.

Q 2. Franklin is famous for another reason. What is it?

Electric matter

Electricity has been with us since the beginning of the Universe, some 15 billion years ago.

Scientists believe that when the Universe was only about three seconds old, the basic particles that make up all matter came into being. They are called protons, neutrons, and electrons. Protons and electrons possess electricity. We also say they have an electric charge. Neutrons have no electric charge.

A proton has exactly the same amount of electric charge as an electron. But it has a different kind of electricity. A proton has what we call a positive (+) electric charge; an electron has a negative (−) electric charge.

Building atoms

All the matter in the Universe is made up of protons, neutrons, and electrons. They are combined together in the form of larger particles we call atoms. But atoms are still very, very tiny. It would take millions upon millions of them to cover the period at the end of this sentence.

The diagram below gives you an idea of how an atom is made up. A bundle of protons and neutrons forms a core, or nucleus, at the center. The electrons circle around the nucleus. Most of the atom, however, is made up of empty space.

It is a basic law of electricity that positive and negative electric charges attract one another. It is the attraction between the positively charged protons and negatively charged electrons that holds the atom together.

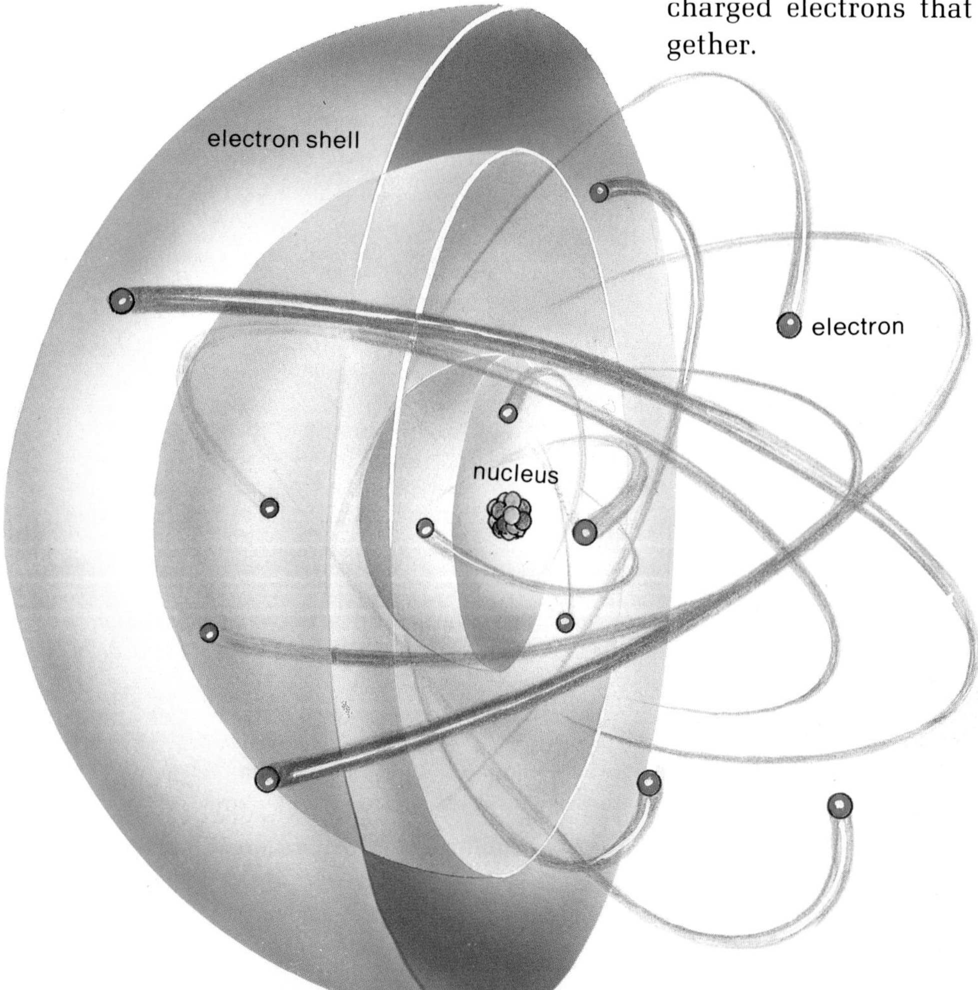

◄ A simple way of looking at the atom. Two main particles, protons and neutrons, are found at the center, forming the nucleus. Electrons circle around the nucleus. They form into distinct groups, circling at different distances from the nucleus in so-called electron shells.

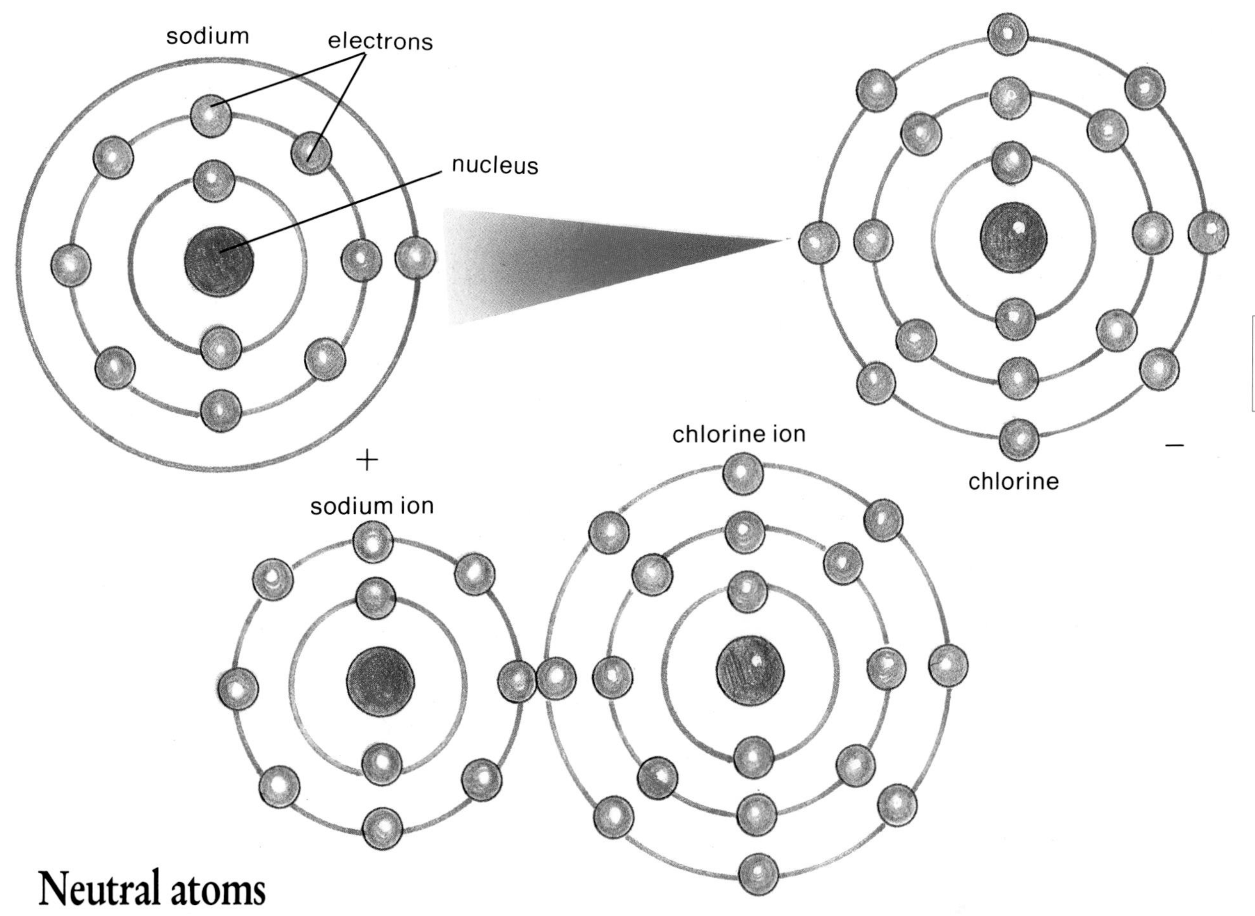

Neutral atoms

There are equal numbers of protons and electrons in an atom. The amount of positive charge of the protons exactly balances the amount of negative charge of the electrons. This makes the atom as a whole electrically neutral – it has no overall electric charge.

Q **1.** Why is the neutron well named?

Combinations

Electrical forces come into play when some atoms combine with one another. When sodium combines with chlorine to form the compound sodium chloride, electrons are transferred from the sodium to the chlorine atoms. This leaves the sodium atoms with a negative electric charge. The attraction between the two kinds of ions (charged atoms) holds the compound sodium chloride together.

Q **2.** Where in nature can we find sodium chloride?

▼ The compound sodium chloride forms crystals, in which the sodium and chlorine ions are arranged alternately in a cubic formation.

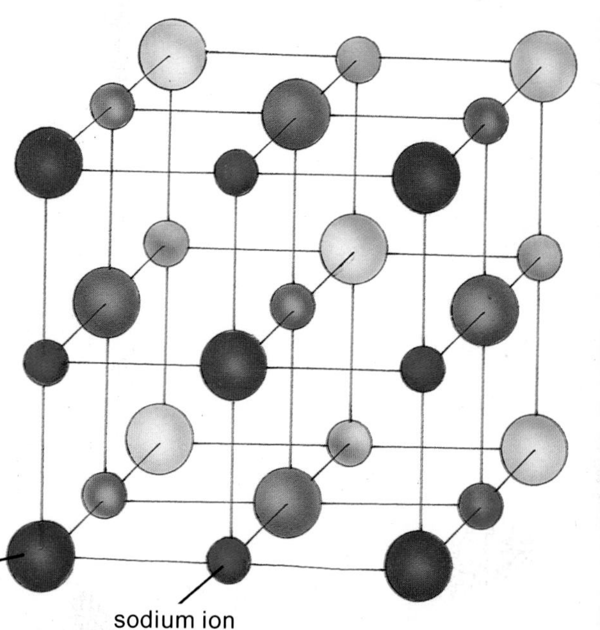

▲ Your sweater isn't sticky; balloons aren't sticky. But balloons will stick to your sweater if you rub them against it first. Your sweater and the balloons become electrically charged and attract one another.

Standing charges

All matter is held together by the attraction between positive and negative electric charges (see pages 10/11). But overall matter is electrically neutral, because the number of positive charges equals the number of negative charges.

A balloon and a cloth, for example, are electrically neutral, having equal numbers of positive and negative charges (see below). But when you rub them together, negatively charged electrons from atoms in the cloth transfer to the balloon.

The balloon now has extra electrons and thus an extra negative electric charge. On the other hand, the atoms in the cloth that have lost their electrons end up with a positive electric charge. The charges tend to stay where they are, or remain static.

Q What else will your balloons stick to after you have rubbed them? The walls, the ceiling, windows, doors, a car?

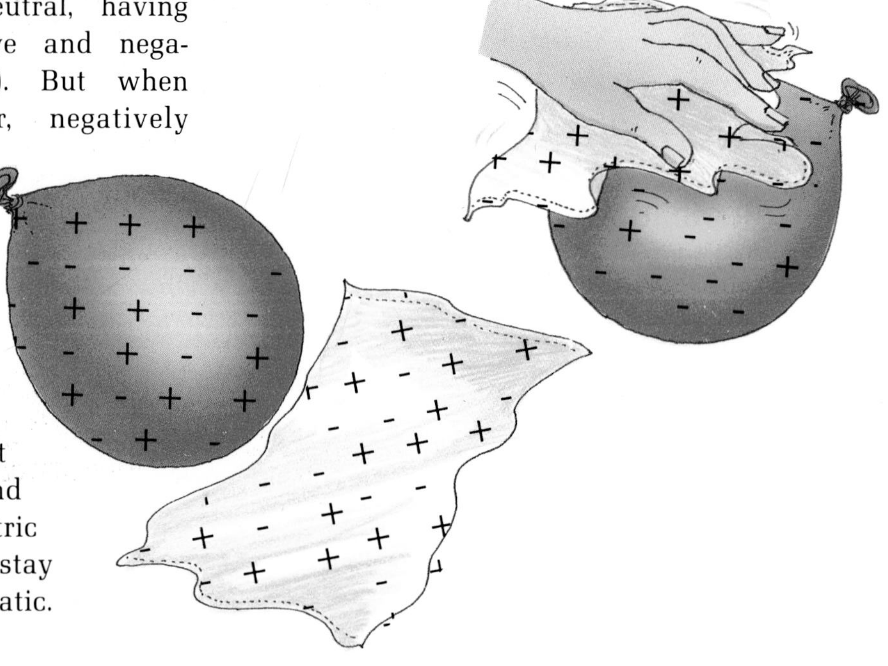

▼ Rubbing a balloon with a cloth charges them both with static electricity.

We therefore call this kind of electricity static electricity.

Q 1. The electrons needed energy to travel from the cloth to the balloon. Where did this energy come from?

Electric laws

When you hold the cloth near the balloon you have been rubbing, the two will be attracted to each other. The negatively charged balloon and the positively charged cloth attract one another.

Now rub two balloons with a cloth, and then dangle them side by side. What happens? You will find that they push away, or repel each other. By rubbing them, you have given them both a negative electric charge. So you have shown that when two negative charges are brought together, they repel one another. You would also find that two objects with a positive electric charge would repel each other.

From these simple experiments, we can now state two basic laws of static electricity: unlike opposite charges attract; like (similar) carges repel.

Hair-raising

Have you ever been really frightened? If you have, you might have felt your hair stand on end.

But there is a less frightening way of getting your hair to do this. Just comb it vigorously with a plastic comb, and then hold the comb above your hair. Make sure your hair is dry.

Q 2. The hairs tend to separate from one another after you have combed them. Why?

INVESTIGATE

Here is a collection of everyday objects. One by one, rub them against a sweater or a piece of cloth and see if they become electrically charged. Tear the paper into little bits. If an object becomes electrically charged, it will pick up the bits of paper. It will also pick up carpet fluff, bits of wool, and so forth. But will it pick up pins? Which objects become electrically charged?

Find out if this statement is correct: Rubbing a metal spoon against your sweater makes it electrically charged so that it will pick up pins.

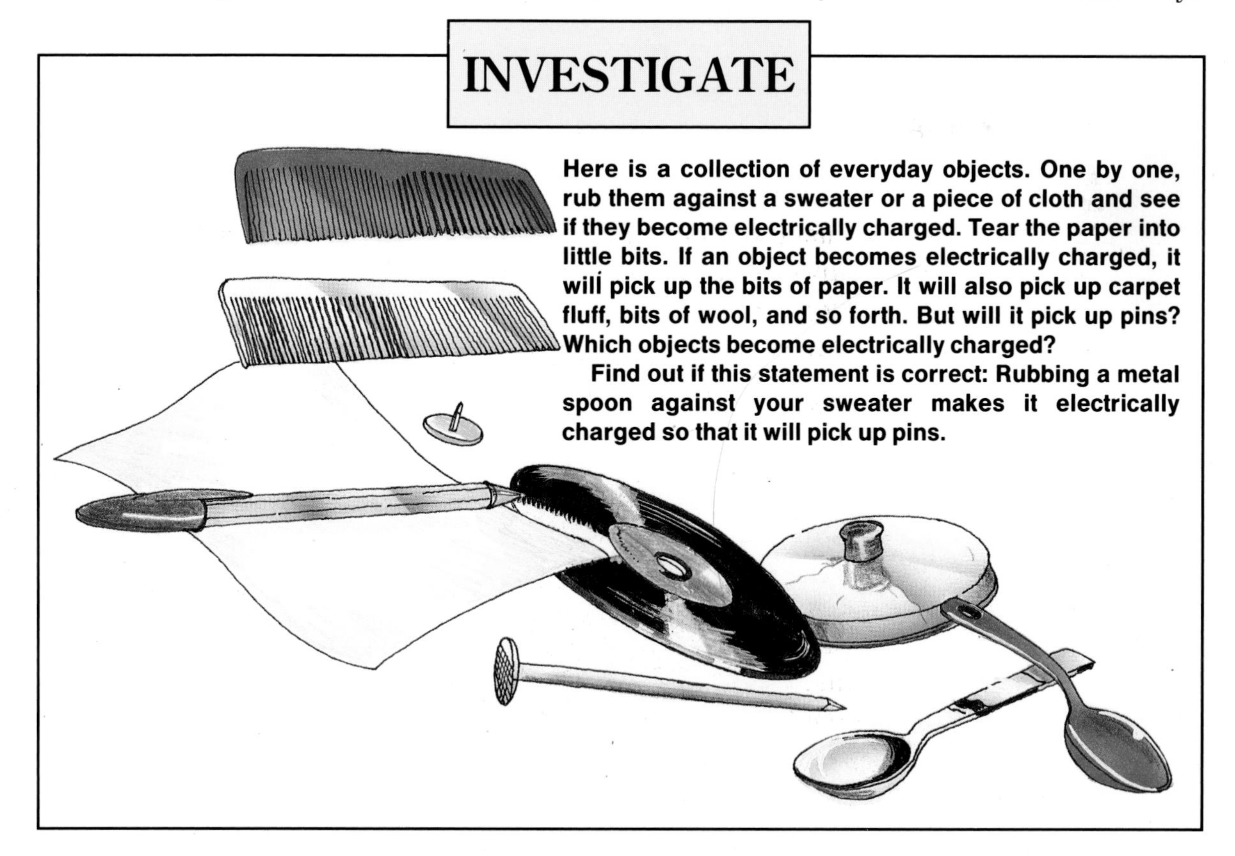

Bright sparks

When your hair is very dry and you comb it with a plastic comb, you can often hear it crackling. You sometimes hear a similar crackle when you pull off your sweater over a shirt or blouse. If you do this in the dark, you may even see tiny flashes of light.

The crackles and flashes come from tiny electric sparks. The action of combing or pulling a sweater over a shirt or blouse sets up static electricity. Sparks occur when this electricity suddenly leaks away, or discharges.

Lightning strike

Much more powerful, and very dangerous, electric sparks occur in nature. We know them better as lightning. In a large thundercloud, little droplets of water and particles of ice race around furiously. They become electrically charged, and these charges build up in different parts of the clouds. For example, positive charges can build up near the top of the cloud, and negative charges near the bottom.

Eventually, a very high electric potential, or "pressure," is created. This is measured in tens of millions of volts. Normally, air prevents electricity in a cloud from leaking away. But in a thundercloud, the pressure is so great that the electricity suddenly discharges. It finds a conducting path either to another cloud or down to the ground.

As the electricity streaks away, it heats the air white-hot and becomes visible as a lightning flash. The air along the conducting path expands suddenly.

Q What does this cause?

◀◀ In a thunderstorm electric charges build up in the thunderclouds. This creates a very high electric "pressure," which is measured in millions of volts. When the "pressure" gets high enough, the electricity discharges to another cloud or to the ground.

Tall trees are often struck by lightning, as they are often the highest points around. High buildings such as churches with spires and skyscrapers are fitted with a lightning conductor, or lightning rod. It provides a conducting path if lightning does strike and helps prevent damage.

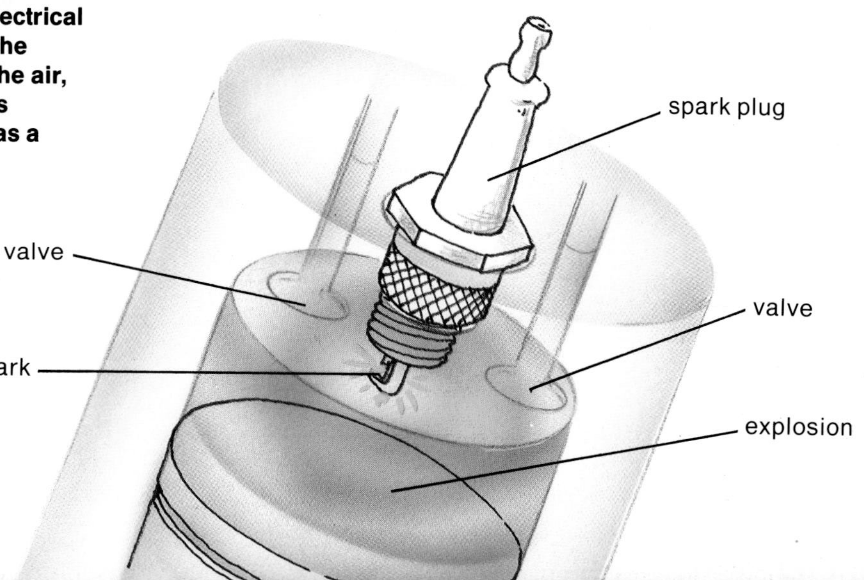

▼ An electric spark is used to explode the fuel mixture in a gasoline engine. The spark is produced when electricity at about 25,000 volts jumps across the gap between the electrodes of the spark plug. This electrical "pressure" overcomes the electrical assistance of the air, and the electricity surges between the electrodes as a spark.

Magnetism

If you spill a box of pins on a rug, picking them up with your fingers can be very painful. But there is an easier way to pick them up – with a magnet. A magnet attracts, or pulls toward it, objects made of iron and steel, such as pins and nails. Ordinary magnets are themselves made of iron or steel. Steel is an iron alloy, or metal mixture.

Do magnets attract other metals as well as iron? Buy one at a hobby shop and investigate for yourself. Will the magnet attract aluminum pans, copper pennies, silver dimes, gold bracelets, or silver charms? And what about other materials – wood, plastics, glass, concrete, and so forth?

You will find that your magnet won't attract any of these metals or other materials. Magnets attract only iron and steel, and the metals cobalt and nickel. You can also make magnets out of cobalt and nickel. Some of the most powerful ones are made out of iron, cobalt, and nickel alloys.

Q Take your magnet and hold it near an empty tin can. Will your magnet pick it up? You'll find that it will. Does this mean that magnets attract tin?

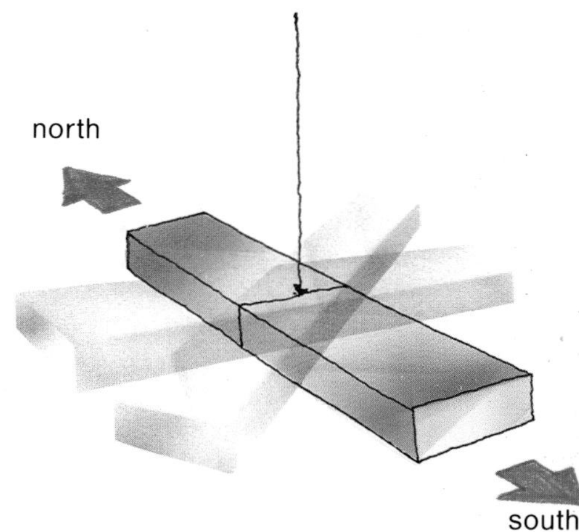

▲ Dangle a bar magnet from a piece of thread. You will find that it will start swinging. When it stops, it will always point in the same direction. One end will point south, the other north. (If no magnetic materials are nearby to attract the magnet).

North poles and south poles

If you push your magnet into a box of pins, you will find

INVESTIGATE

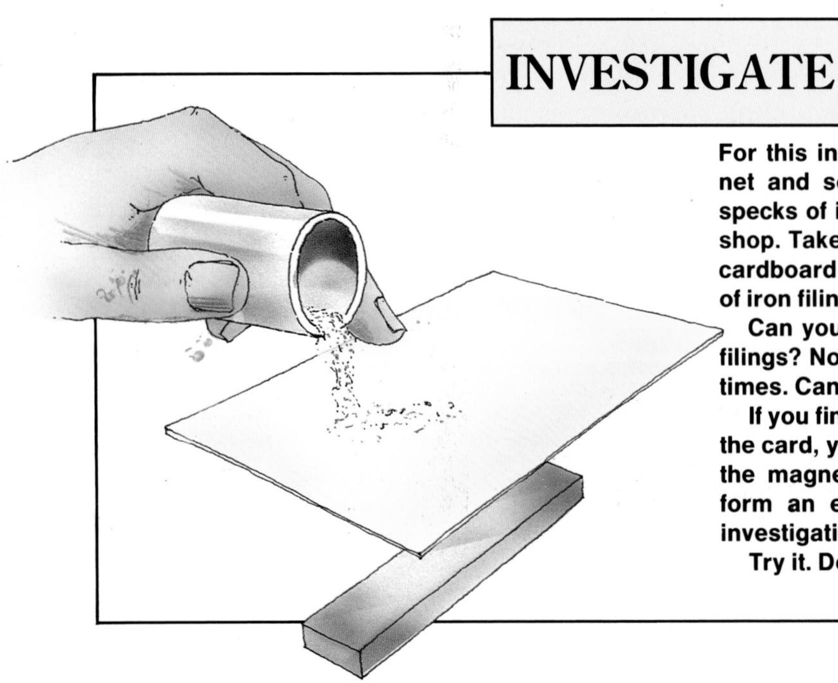

For this investigation you will need a magnet and some iron filings, which are tiny specks of iron. You can buy both at a hobby shop. Take the magnet and place a piece of cardboard on top of it. Sprinkle a thin layer of iron filings onto the cardboard.

Can you see any pattern in the sprinkled filings? Now tap the cardboard lightly a few times. Can you see any pattern now?

If you find that the filings form a pattern on the card, you can't immediately assume that the magnet is responsible. You must perform an experiment. Carry out the same investigation without the magnet in place.

Try it. Do the filings form a pattern?

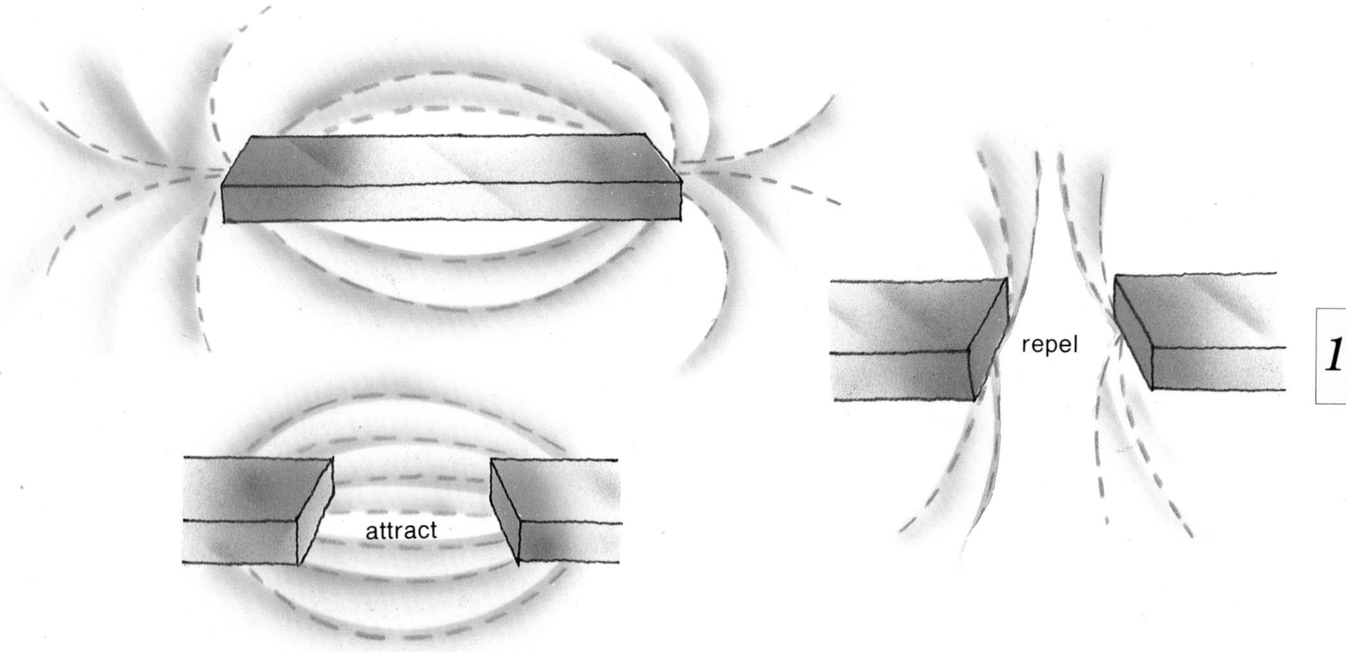

that the pins stick mainly to its ends. This shows that magnetism is concentrated at the ends of a magnet.

If you suspend a magnet from a thread, it will always come to rest pointing north and south, if no magnetic materials are nearby. We call the north-pointing end of a magnet the north pole, and the south-pointing end the south pole.

Bring the poles of two magnets together (see above). You will find that a north pole and a south pole will attract each other. But two north or two south poles will repel each other.

This is similar to what happens when the two different kinds of electric charges, positive and negative, come together (see page 12). We can state a similar law for magnets: unlike poles attract, like poles repel.

If you could see the magnetic forces around a magnet, they would look like the illustration at the top of the page. The lines are called magnetic lines of force. The whole region over which the forces act is called the magnetic field.

When the poles of two magnets are brought close together, their magnetic forces affect one another. With a south and a north pole, the lines of force join up, and the poles attract each other. With two north or two south poles, the lines of force separate, and the poles push against each other.

▼ Information is stored on computer disks in the form of magnetized particles.

Q Other devices found in the home also use magnetized particles for storage. What are they and what do they store?

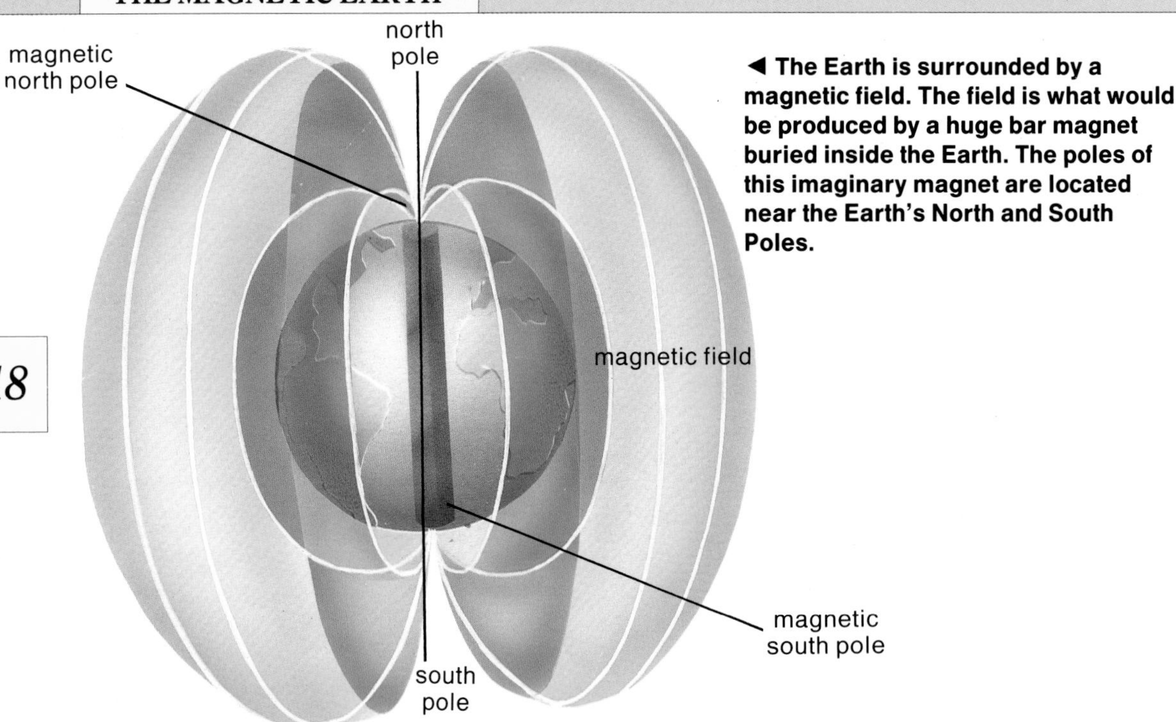

◀ **The Earth is surrounded by a magnetic field. The field is what would be produced by a huge bar magnet buried inside the Earth. The poles of this imaginary magnet are located near the Earth's North and South Poles.**

The magnetic Earth

Suspend a bar magnet on a thread, and it will always come to rest pointing north-south. Why does this happen?

Obviously, the magnet is being affected by another magnetic field – the magnetic field of the Earth. This field has lines of force pointing north-south, and the magnet aligns itself with these lines.

The Earth behaves as if it contains a magnet, with one magnetic pole near the North Pole and the other magnetic pole near the South Pole. In fact there is no such Earth-magnet. Scientists think that the Earth's magnetism may be produced as a result of electric currents flowing through the liquid iron that forms the Earth's outer core.

The magnetic compass

The fact that a magnet always aligns itself with the north-south lines of the Earth's magnetic field makes the compass possible. We use a compass when we want to find directions. We need to know directions when we are reading a map.

▼ **The needle of a magnetic compass always points north-south and is therefore useful for finding direction.**

Q If you lose your compass when you are map-reading at night, how can you still locate north?

► **The Earth's magnetic field extends out to great distances in space. It forms a gigantic magnetic "bubble," which we call the magnetosphere. The "bubble" is distorted because of the solar wind "blowing" from the Sun.**

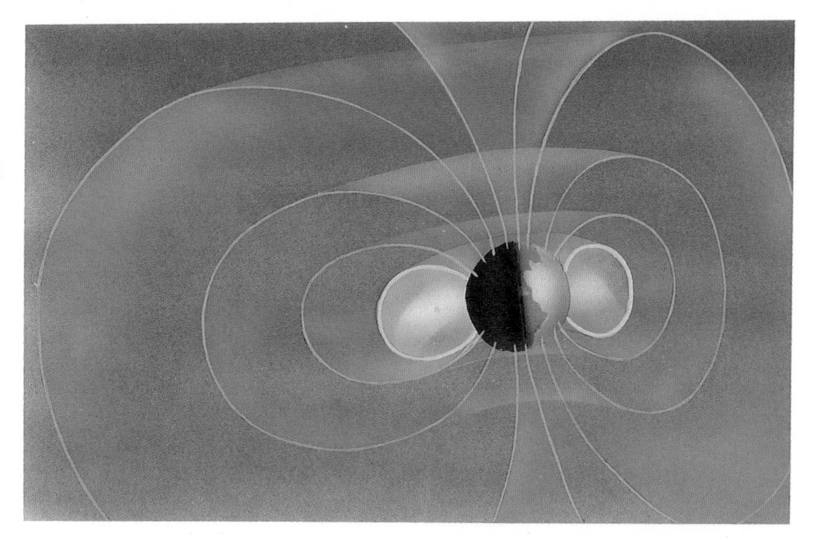

The compass consists of a magnetized needle, mounted on a pivot so that it is free to rotate. Because it is magnetized, it always comes to rest pointing north-south. The scale of the compass shows the other "points," or directions.

A compass does not point exactly north-south, however. This is because the Earth's magnetic poles are not in the same position as the geographic poles; that is, the North and South Poles. So when you are map-reading, you must allow for this difference in position.

The angle between the magnetic and geographic north poles is called the magnetic variation, or declination. It gradually changes year by year and varies from place to place.

Q **1.** If you used a pocket compass on board an ocean liner traveling along the Equator, in which direction would the needle point?

◄ **A brilliant display of the Southern Lights in the far Southern Hemisphere, pictured from the space shuttle. Such displays are called the Northern Lights in the Northern Hemisphere. They take place when charged atomic particles spiral into the atmosphere along magnetic lines of force and collide with air particles.**

Q 2. Do you know the scientific name for such displays?

2
Current Affairs

◀ **Power lines snake across the countryside, supported by tall steel pylons. The lines hang from the arms on insulators.**

When you comb your hair with a plastic comb, you give the comb and your hair an electric charge. This happens because negatively charged electrons move from the comb to the hair and stay there for a while. If you comb your hair with a metal comb, neither the comb nor your hair become electrically charged.

Electrons can flow easily through metals, and it is a flow of electrons that we call an electric current. Current electricity differs from the static electricity of electric charges, which stays where it is.

Current electricity is the kind of electricity we are most familiar with. It comes from batteries and from the electricity supplied to our homes through cables. Electricity for our homes is produced in power stations, using machines that rely on the principles of electromagnetism. Electromagnetism is a term which refers to an interesting relationship between electricity and magnetism.

▼ **Batteries provide the electricity that powers flashlights. The batteries are connected in a circuit (path) with a light bulb. The electricity passes through the wire in the bulb and makes it so hot that it gives out light. A switch is included in the circuit to turn the electric current on and off.**

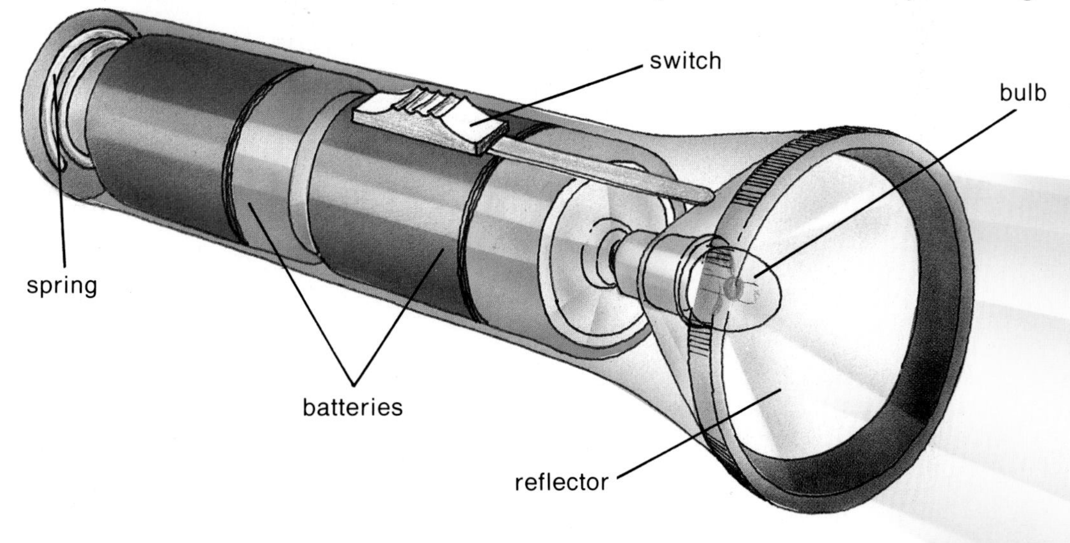

Animal electricity

One of the most interesting stories in the history of electricity concerns an Italian professor from the University of Bologna named Luigi Galvani. In the 1780s he began electrical experiments with frogs' legs. He found he could make the leg muscles of a dead frog twitch by giving them an electric charge.

He later found that the legs twitched of their own accord when they were hung from a copper hook on an iron frame and touched the frame. He therefore concluded that the muscles in the frog's legs produced electricity, "animal electricity." Experiments with this new, exciting electrical "force" became all the rage.

Galvanism

But other scientists were skeptical about the existence of animal electricity. One was Galvani's friend Professor Alessandro Volta of the University of Pavia. Volta reasoned that the electricity was produced, not by the

▼ Some of the apparatus used in an electrical laboratory of the 18th century. Frogs' legs are everywhere! Some early experimenters thought that they were a source of electricity. The disk machine was used to generate static electricity.

frog's muscles, but by the different metals they were in contact with – copper and iron.

Volta was right, and this led him in 1800 to produce the first electric battery. By then, Galvani had been dead two years. But his name lived on in the term "galvanism," which Volta coined for the new kind of flowing electricity his battery produced.

Nervous impulses

Galvani may have been wrong about "animal electricity," but he showed that electricity causes animal muscles to contract. In fact, scientists have shown that messages are sent between the brain and the rest of the body by tiny electrical impulses. The pathway for these impulses is the central nervous system.

Special cells called nerve cells, or neurons, carry the impulses. The body tissues we call nerves consist of many long neurons bundled together, somewhat like a telephone cable consists of a bundle of separate wires.

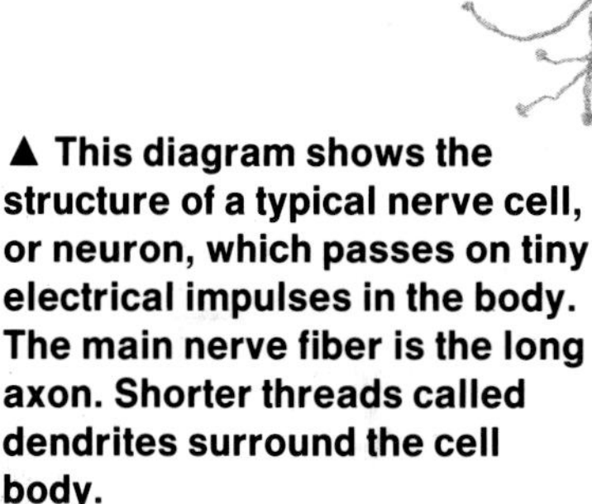

▲ This diagram shows the structure of a typical nerve cell, or neuron, which passes on tiny electrical impulses in the body. The main nerve fiber is the long axon. Shorter threads called dendrites surround the cell body.

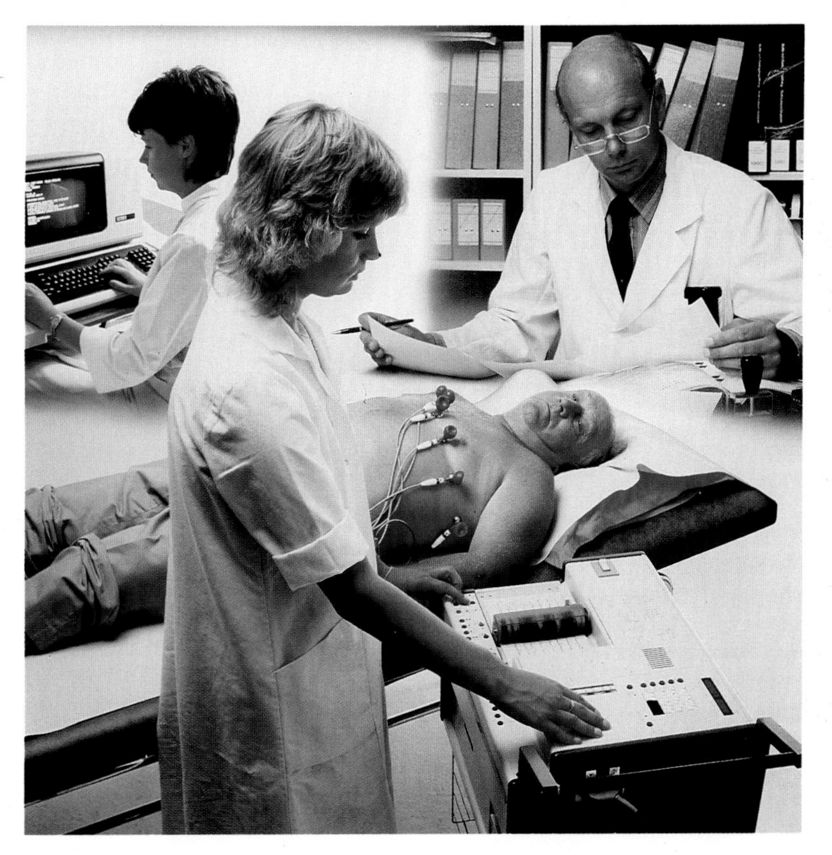

◀ A patient is wired up for an EKG, or electrocardiograph. This instrument records the electrical activity of the heart. From the EKG abnormalities in the heart's rhythm may be detected.

Currents and circuits

The ordinary electricity we use in our everyday lives has to do with electrons flowing, usually through wires. A flow of electrons is an electric current. For many uses, as in flashlights and pocket radios, batteries provide the electrical "pressure" that makes the current flow.

In a flashlight, the current makes a light bulb light up. For this to happen, there must be a continuous path from one terminal of the battery to the bulb and then back to the other battery terminal. We call such an electrical path a circuit. The diagram below shows the features of a typical circuit.

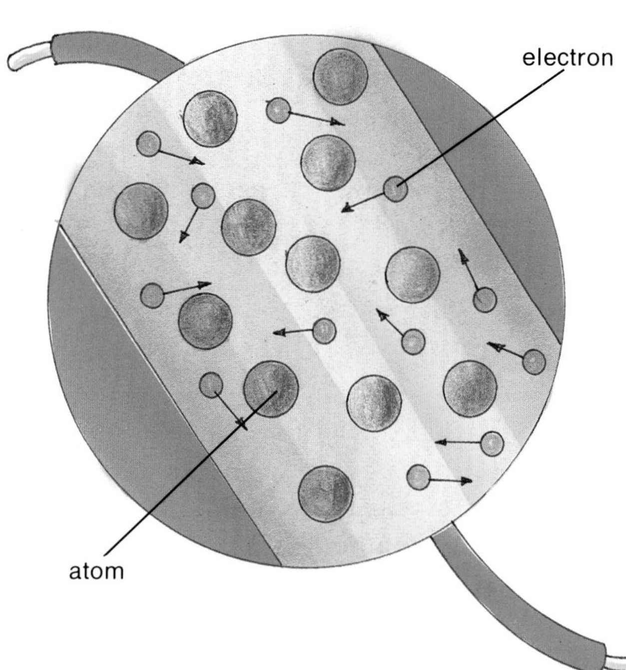

▲ Electrons are free to wander among the atoms in a metal; they usually wander at random.

Good conductors

In electric circuits, copper wires are usually used to make connections. This is because copper conducts, or passes on, electricity well. Most metals, in fact, are good conductors. This is because the electrons in their atoms are free to wander around. When a battery provides the "pressure," the electrons start moving, setting up an electric current.

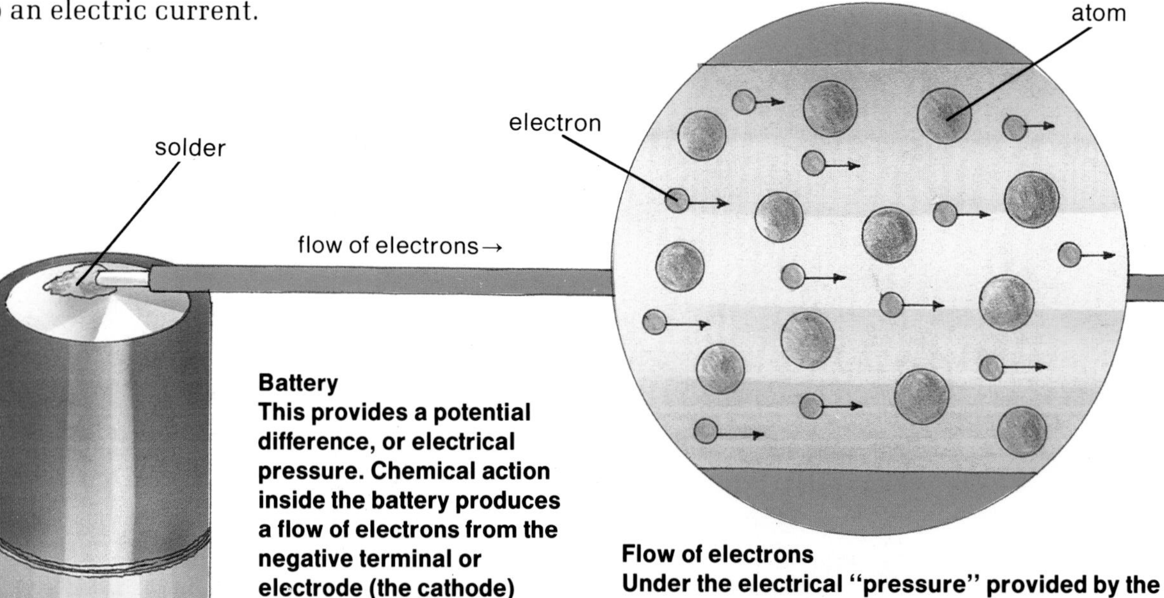

Battery
This provides a potential difference, or electrical pressure. Chemical action inside the battery produces a flow of electrons from the negative terminal or electrode (the cathode) around the circuit to the positive terminal or electrode (the anode).

Flow of electrons
Under the electrical "pressure" provided by the battery, the electrons wandering among the atoms in the metal wire now all move in the same direction, toward the positive terminal.

The electrical "pressure" of a battery is measured in units of volts, honoring the person who developed the first battery, Alessandro Volta (see page 28). Electric current is measured in units called amperes, or amps, named after the French electrical pioneer Andre Ampere.

When electricity flows through any conductor, it experiences a certain amount of resistance, just like water flowing through a pipe. Electrical resistance in a circuit is measured in units called ohms, named after the German physicist Georg Ohm. It was he who first established a simple relationship in an electric circuit, that current equals voltage divided by resistance. This is known as Ohm's law.

Q Think of the similarity between electric current flowing through a wire and water flowing through a pipe. Which would have the greater resistance: a long wire or a short wire; a thin wire or a thick one?

Confusion!

In an electric circuit, the flowing electrons that make up an electric current travel from the negative battery terminal to the positive. Many people think that electric current travels from positive to negative!

Remember, remember

Here is a rhyme to help you remember a form of Ohm's Law:

The Resistance of lamps
is Volts over Amps.

In other words, the resistance in an electrical circuit is the voltage divided by the current.

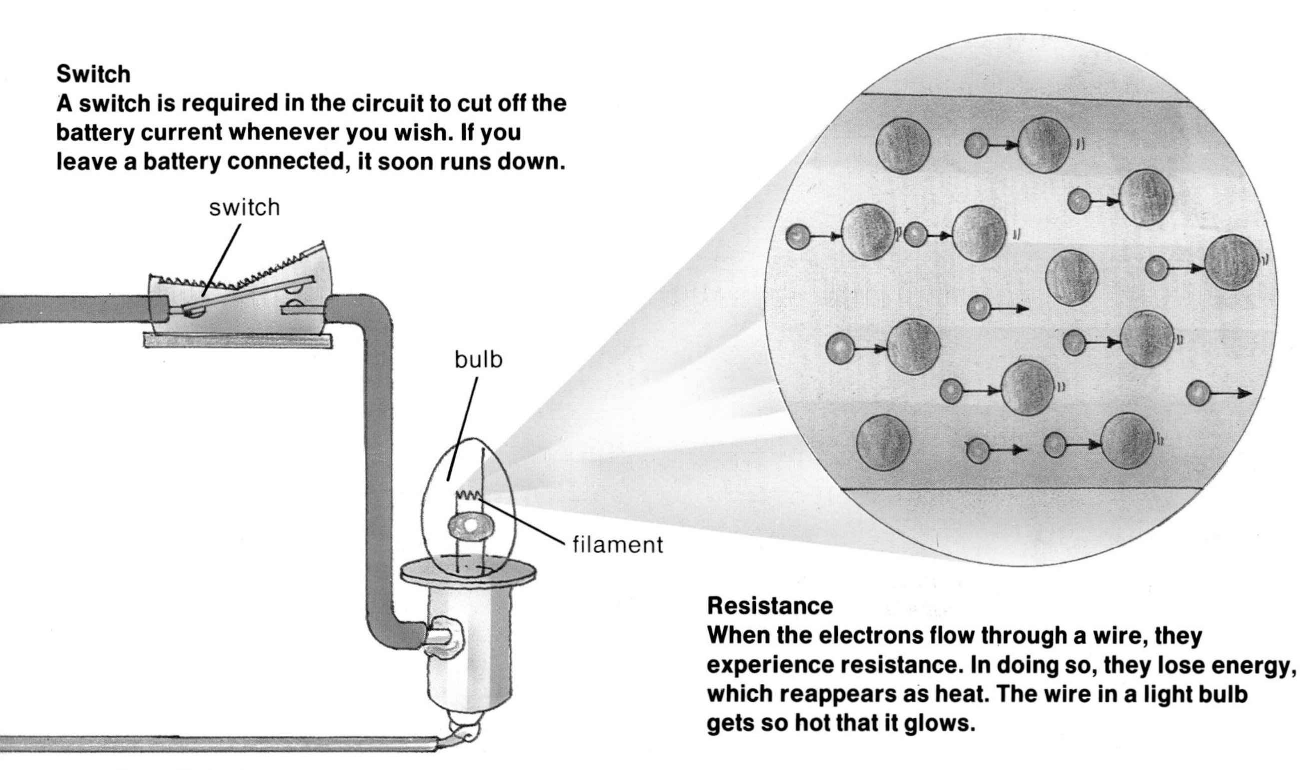

Switch
A switch is required in the circuit to cut off the battery current whenever you wish. If you leave a battery connected, it soon runs down.

Resistance
When the electrons flow through a wire, they experience resistance. In doing so, they lose energy, which reappears as heat. The wire in a light bulb gets so hot that it glows.

More about conductors

As mentioned earlier, most metals are good conductors of electricity. Copper is an excellent conductor and is the metal most widely used in electrical devices. Silver is the best conductor of all, but it is too expensive to be used on a large scale.

Gold is also an excellent conductor, now used widely in the form of very fine wire to make the delicate connections required in microchips.

Q **1.** Gold has two main properties that make it better-suited for such a delicate use than most other metals. Can you think of what they might be?

Nonmetals

In general, nonmetals do not conduct electricity. The electrons in the atoms in these substances cannot "wander" around as they can in metals. We call substances that do not conduct electricity insulators. Particularly good insulators include glass, ceramics, rubber, and plastics. The air is a good insulator, too.

However, there is a notable exception to the rule: the element carbon. In the form of graphite, carbon is a good conductor. Carbon rods are used for the positive electrodes in ordinary dry cells. Diamond, the other main natural form of carbon, is a poor conductor.

Q **2.** What, however, is the main difference between graphite and diamond?

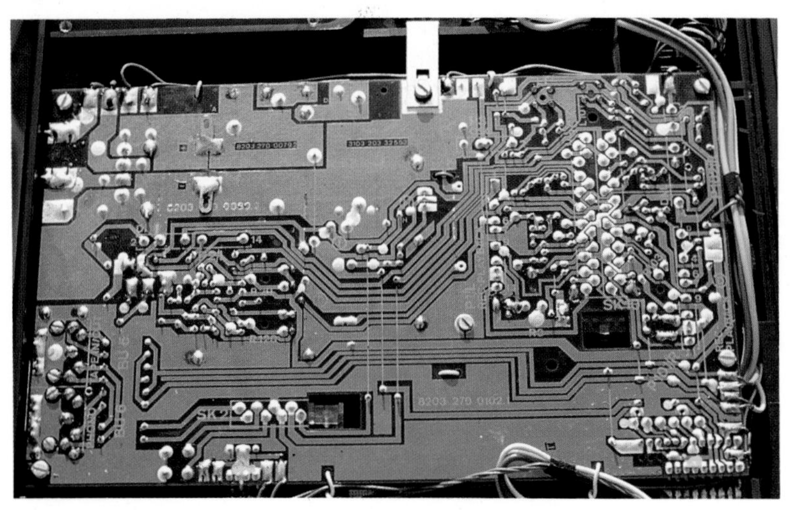

◀ The underside of a printed-circuit board of a radio. The circuit consists of a film of copper, laid down on an insulated board by a photo-printing technique. The silver blobs show where wires from the electronic components in the radio are held in place in the circuit by solder.

Liquid conductors

Some nonmetals do not conduct electricity when they are solids. But they do conduct electricity when you melt them or dissolve them. We call these conducting liquids electrolytes.

The chemical substances we call salts (which include ordinary table salt) are examples. Molten salts and salts in solution conduct electricity well. This is because they are made up, even in the solid state, of ions. When they are melted or dissolved in solutions, the ions are free to move and conduct electricity.

Q Can you remember what ions are? Why can they conduct electricity?

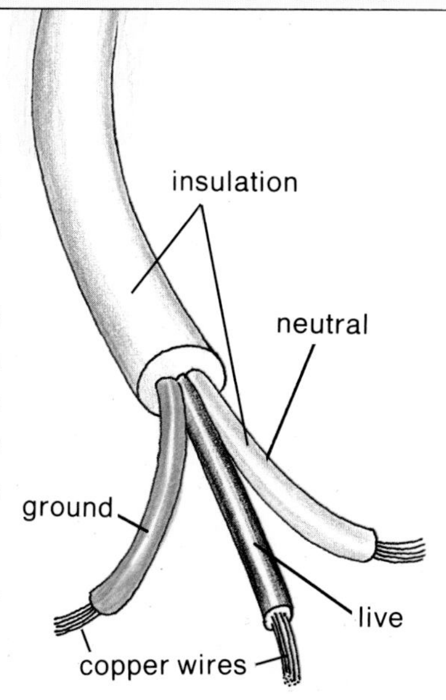

▲ **Electric cables made up of three wires are used for wiring power sockets. Each wire is insulated by a plastic sheath, preventing electricity from passing between them.**

Semiconductors

Some substances fall between conductors and insulators. Called semiconductors, they conduct a little electricity under certain conditions. The best-known semiconductor is silicon, an element closely related to carbon. This is the semiconductor most widely used in the production of microchips (see page 50).

Silicon conducts electricity a little when it contains certain impurities. The presence of impurities adds a few extra electrons to the atomic structure. Movement of these electrons gives rise to an electric current.

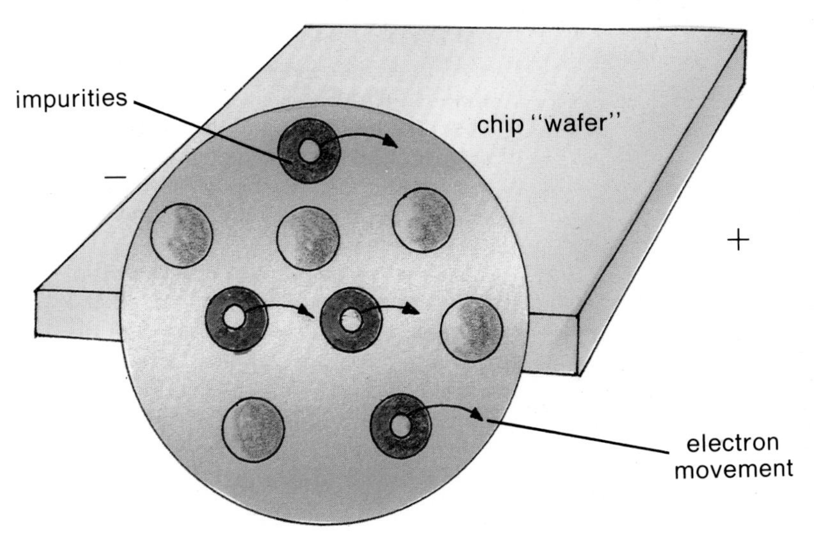

◀ **In one kind of semiconductor, impurities add a few spare electrons. Under suitable conditions, these electrons can move, allowing a trickle of electric current to flow.**

Cells and batteries

The electricity that powers our flashlights, pocket radios, Walkmans, calculators, hearing aids, and other portable electrical devices is supplied by one or more batteries, or electric cells, as they are technically called.

The first battery was made in 1800 by the Italian Alessandro Volta, the scientist who debunked Galvani's idea of "animal electricity" (see page 22). His battery was called the voltaic pile. It consisted of alternate discs of silver and zinc, with sheets of pasteboard soaked in salt water sandwiched in between. This pile was the first device to deliver a steady supply of electricity.

Simple cells

Another early electricity provider was the so-called simple cell. A simple cell consists of a copper (or carbon) rod and a zinc rod dipping into a solution of sulfuric acid. Let us look at the action of this cell, which is typical of all ordinary electric cells.

The metal rods are known as electrodes. The sulfuric acid acts as a conducting liquid, or electrolyte. The chemical action between the electrodes and the electrolyte produces the electricity.

INVESTIGATE

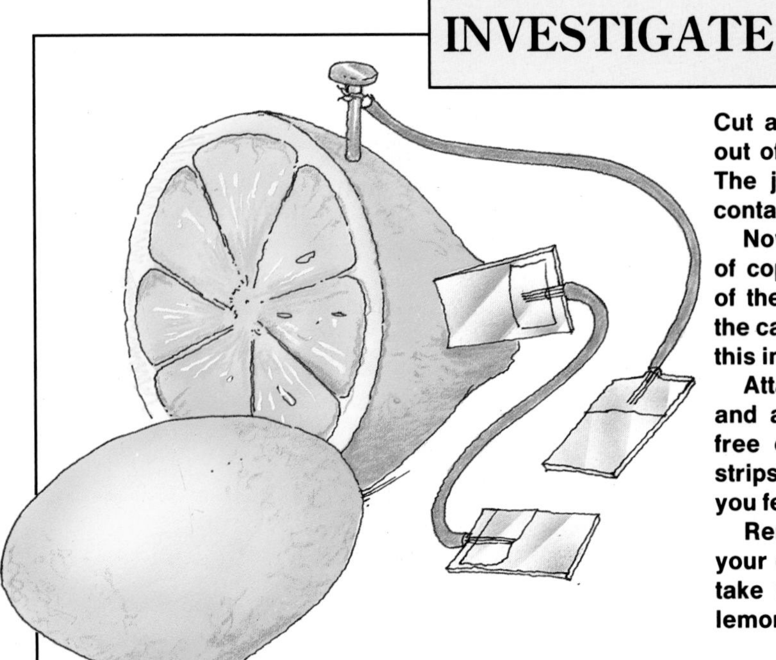

Cut a lemon in half. Squeeze the juice out of one half into a glass and taste it. The juice tastes very sour, because it contains ... What?

Now take a copper nail, or other piece of copper, and stick it in the other half of the lemon. Cut a strip of metal from the casing of an old dry battery, and stick this into the lemon, too.

Attach wires to the nail and the strip, and attach strips of kitchen foil to the free ends of the wires. Place the foil strips against your tongue. What can you feel?

Remember to include a control in your experiment. Repeat the above, but take the nail and the strip out of the lemon first.

The most common battery used today is the so-called dry cell, which has its chemicals sealed inside a casing. Like the simple cell, it is a primary cell, which means it can be used only once. When its chemicals have run out, it stops working.

Other batteries, called secondary cells, can be used more than once. They can be recharged with electricity, which restores their original condition and allows them to produce electricity once again. The lead-acid car battery is made up of secondary cells.

nickel plated steel case
zinc powder
electrolyte
mercuric oxide and graphite

▲ This type of battery is widely used for powering digital watches and pocket calculators. It is a mercury battery. It has a positive electrode of mercuric oxide and graphite and a negative electrode of powdered zinc. The electrolyte is a strong alkaline solution.

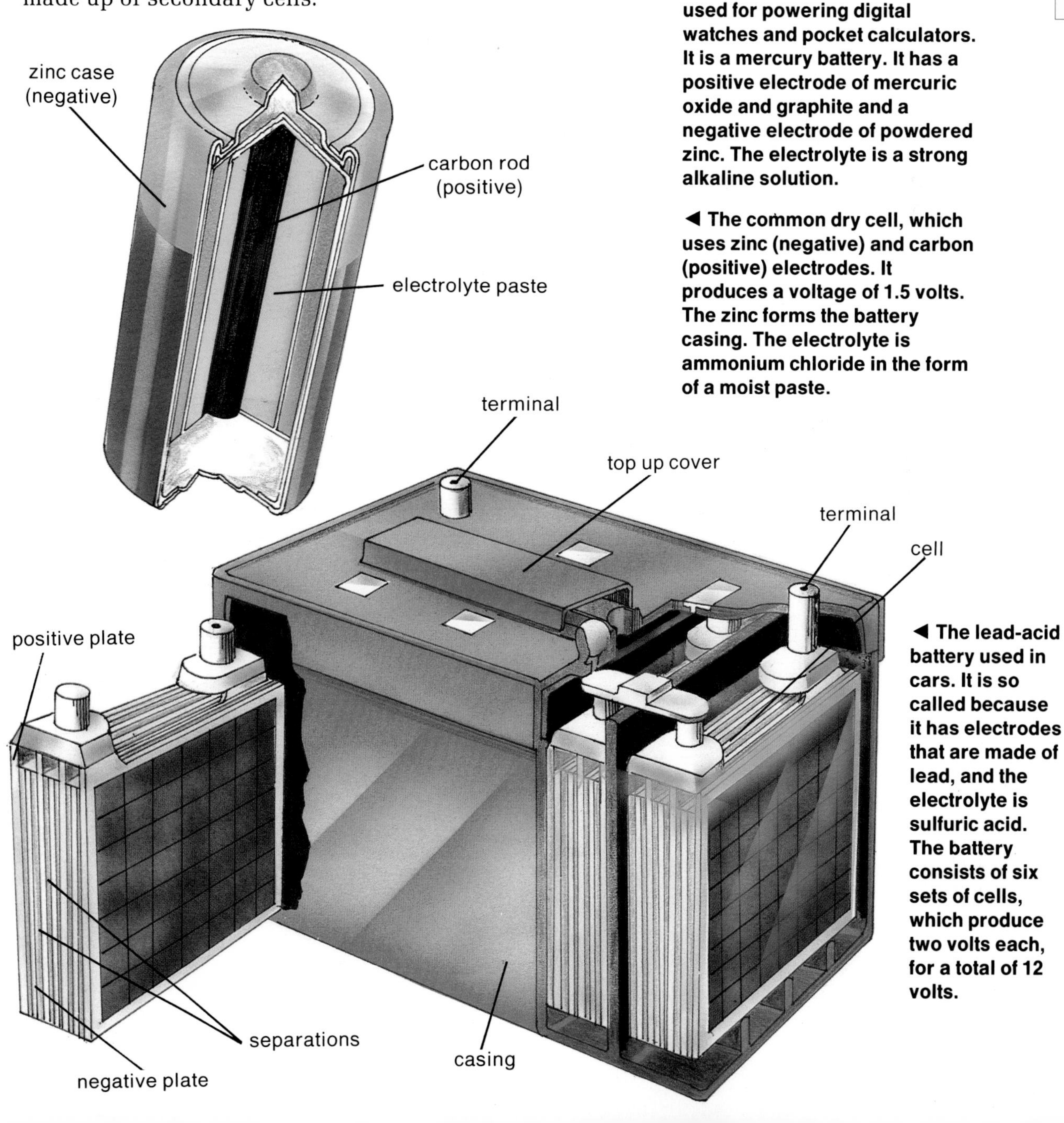

◀ The common dry cell, which uses zinc (negative) and carbon (positive) electrodes. It produces a voltage of 1.5 volts. The zinc forms the battery casing. The electrolyte is ammonium chloride in the form of a moist paste.

◀ The lead-acid battery used in cars. It is so called because it has electrodes that are made of lead, and the electrolyte is sulfuric acid. The battery consists of six sets of cells, which produce two volts each, for a total of 12 volts.

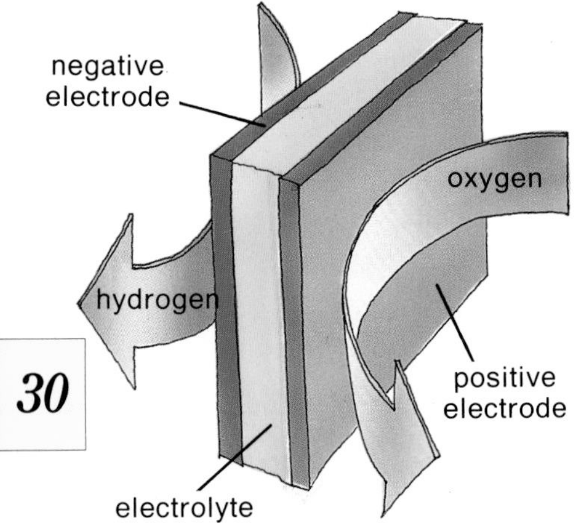

▲ The principle of the fuel cell, a device invented by the British engineer Francis Bacon in 1959. The basic chemical reaction is between hydrogen and oxygen supplied to the electrodes in the form of a gas. Water is produced as a by-product.

Q 2. The fuel cells on the space shuttle produce about 2 gallons (7.5 liters) of water an hour. How much water is produced during a 5-day mission?

Alternative electricity

The battery is by far the most common device for producing electricity. But several other kinds of devices have been developed for providing electric current in circumstances when ordinary batteries will not do.

Some, like the fuel cell, work in a similar way to the ordinary battery. That is, they produce electricity as a result of chemical action between the substances making up the electrodes and the electrolyte.

The fuel cell makes use of the simple chemical reaction between hydrogen and oxygen to form water. It is used on the space shuttle, for example. Fuel cells are highly efficient and can produce electricity indefinitely, as long as they are supplied with fuel.

Q 1. Why are fuel cells particularly useful for supplying power on piloted space missions?

Cells in space

Other alternative sources of electricity produce electric current, not by chemistry, but in quite a different way. Solar cells, for example, harness the energy in sunlight and convert it into electricity.

Shock tactics

Electrophorus electricus is a perfect name for a shocking animal. We know it better as the electric eel. It lives in the Amazon and Orinoco rivers of South America. When the eel becomes excited, special muscles in its body produce electricity in brief pulses, but at voltages of 300 volts or more. The electric shocks are strong enough to stun its prey. Some can be powerful enough to stun a person or even a horse standing in the water.

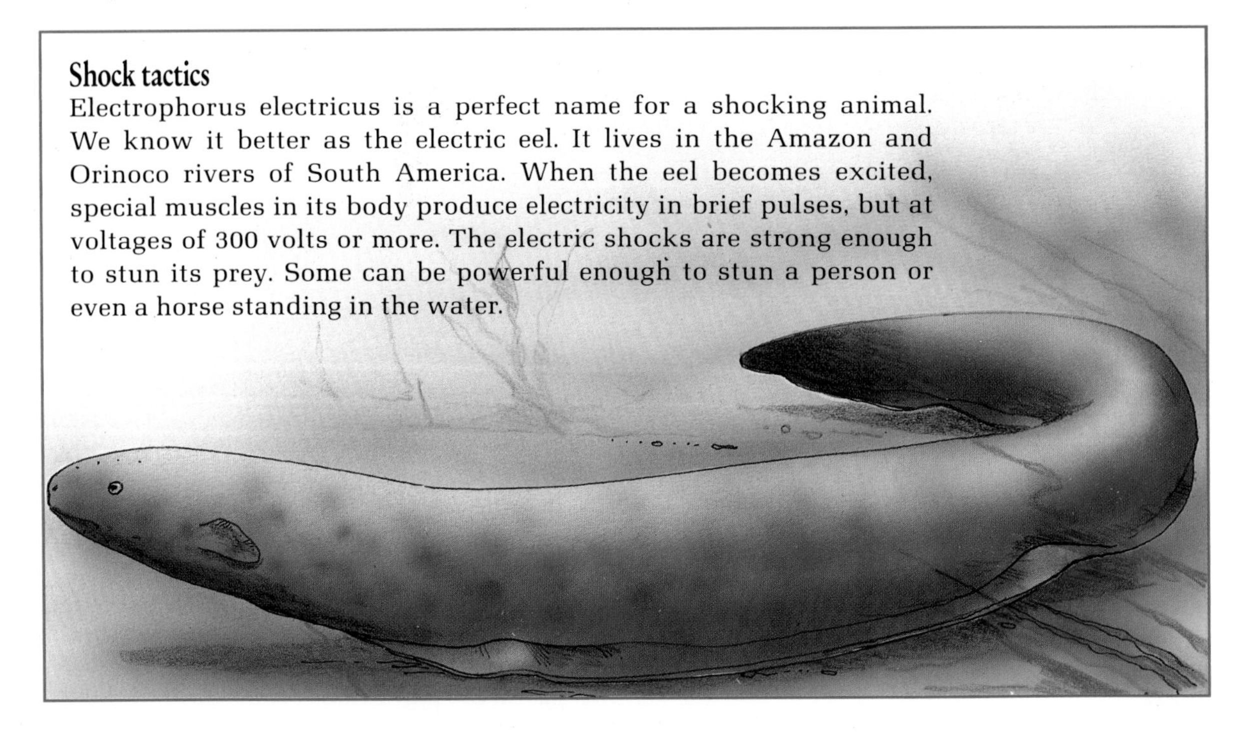

► A scientist at Sandia National Laboratories near Albuquerque, New Mexico, checks the arrays of solar cells, which are being tested there.

Solar cells are made of layers of doped silicon, the material used to make microchips. Electrons start flowing in the cells when sunlight falls on them. They are technically termed photovoltaic ("light-voltage") cells.

Mostly solar cells are used for powering space satellites. They are not used on a large scale on Earth because they are too expensive.

Q Why can't much cheaper batteries be used to power satellites?

► The automatic scientific station set up on the Moon by the Apollo 16 astronauts. At the center is the power source for the instruments, a nuclear "battery." Nuclear batteries contain devices called thermocouples. They convert into electricity the heat given off by a radioactive substance such as plutonium. They are technically called radio isotope thermoelectric generators (RTGs).

Electric chemistry

Electrochemistry is the field of science that deals with the relationship between electricity and chemistry.

We have already seen how chemical reactions taking place inside batteries produce electricity. But the opposite also occurs. Electricity can make chemical reactions take place. This happens when you pass electricity through electrolytes. These are substances that conduct electricity when they are in a solution or when they are melted.

Passing electricity through electrolytes splits up the electrolytes. This process is called electrolysis. For example, when you pass electricity through water, you split it up into its constituent elements, hydrogen and oxygen. Electrolysis is a very important process in the chemical industry.

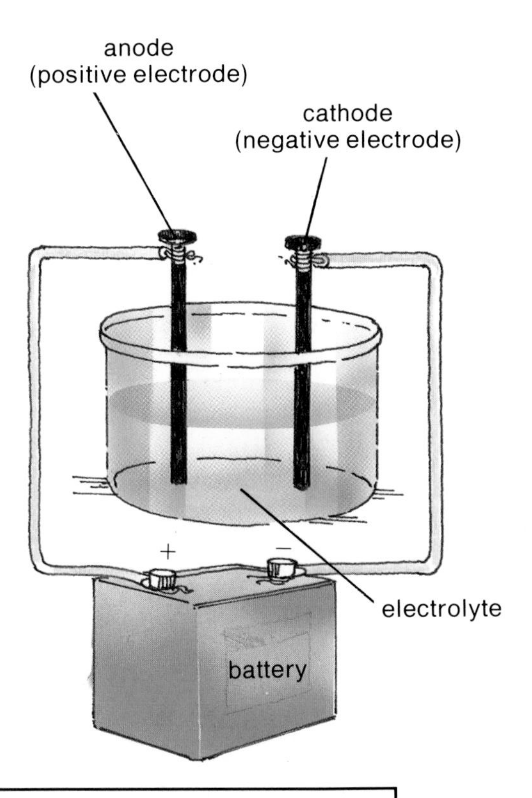

The electrolytic cell

The diagram (right) shows the features of a "cell" used for electrolysis. Electricity is fed into it via an anode (positive electrode) and cathode (negative electrode).

Feeding electric current to the electrodes brings about the break-up, or decomposition, of the substance in be-

INVESTIGATE

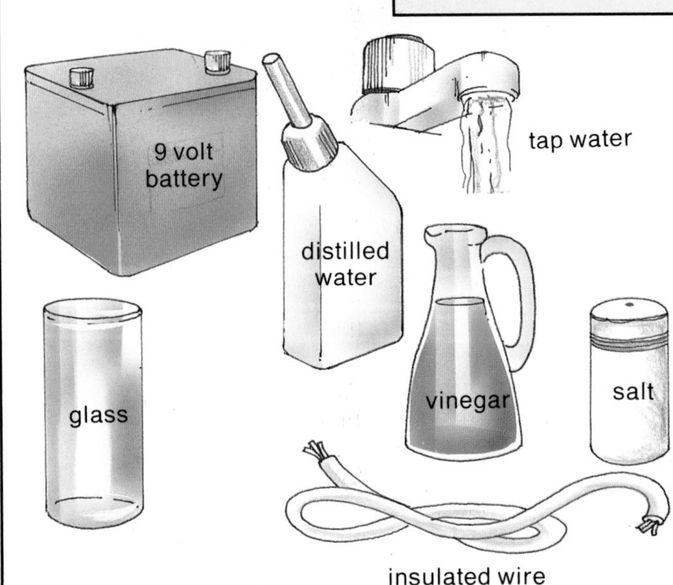

For this investigation, you will need the items shown in the picture.

Pour some distilled water into the glass, swirl it around thoroughly, and then throw it away. Pour in some more distilled water. Connect two lengths of wire to the battery, then dip the free ends of the wire into the water. Watch for a minute. Does anything happen?

Now add a teaspoonful of vinegar to the water. Does anything happen now?

Rinse out the glass thoroughly with distilled water and repeat the experiment, but this time add a spoonful of salt to the water. Do you get the same results?

What happens if you use tap water instead of distilled water?

tween – the electrolyte. Electrolytes split up into ions (charged atoms) when they are dissolved in water or when they are molten (see page 27). In an electrolytic cell, the positive ions are attracted to the cathode, and the negative ions to the anode.

In the electrolysis of molten sodium chloride, for example, the positive sodium ions travel to the cathode, and sodium metal is deposited there. This is the way, in fact, that sodium is usually made.

And in the electrolysis of a silver solution, positive silver ions are attracted to the cathode, and silver metal is deposited there. This process is used to coat objects with silver and is an example of electroplating.

INVESTIGATE

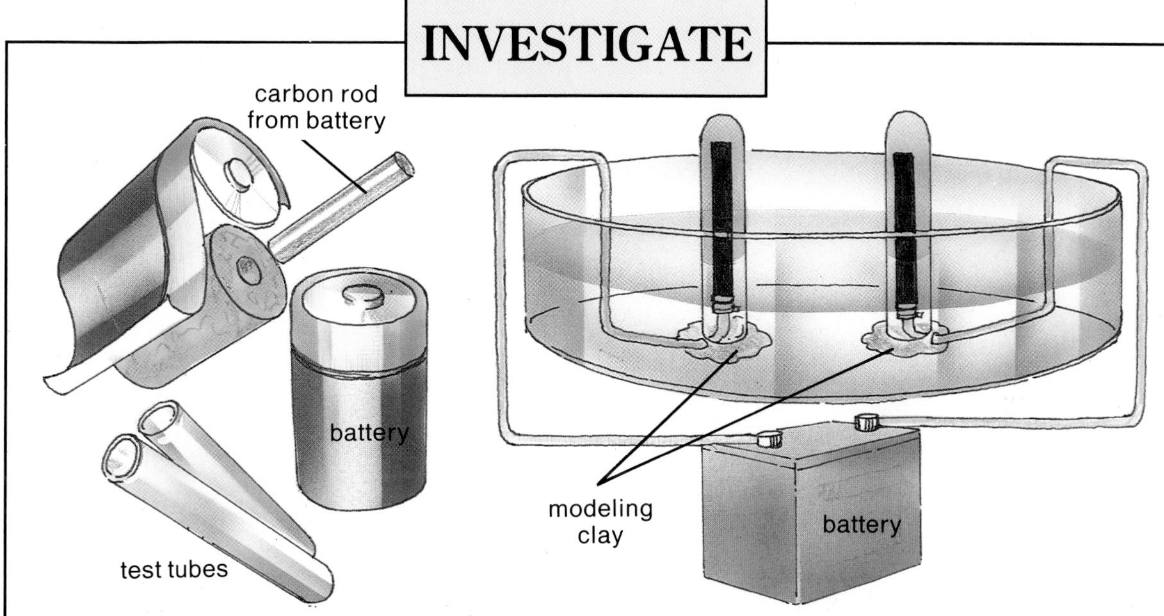

Set up the apparatus shown above, in which you can carry out the electrolysis of water.

DO NOT ATTEMPT THIS INVESTIGATION WITH ELECTRICITY FROM AN ELECTRICAL OUTLET – IT IS DANGEROUS.

For the electrodes, use the carbon rods from old dry batteries. Stand them in modeling clay. Fill the test tubes with water and upend them over the rods. Connect them and you will see gas bubbles form around the rods. The gas will collect at the ends of the tubes and force out the water.

More gas collects in one tube than the other. To which battery terminal (+ or –) is the wire to this tube connected?

In this investigation you split water into its two components, hydrogen and oxygen. Both are colorless and odorless gases. So how can you tell which gas is in which tube?

Wait until you have a tube full of gas, then test it. Hold a thin lighted candle near the mouth of the tube. If it goes "plop," this tells you that it must be ... What?

If you don't get a "plop," blow out the candle, but leave it still smoldering. Now put the candle into the tube. What happens?

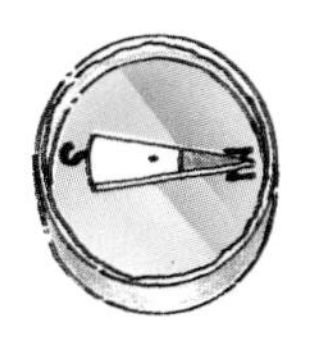

▲ The needle of a compass placed next to a length of wire points north.

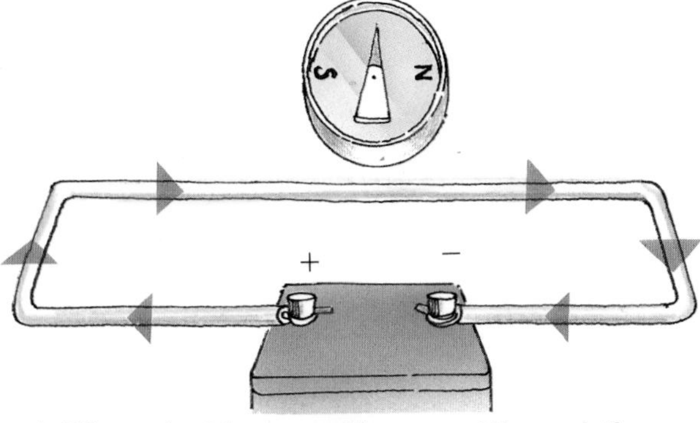

▲ When electric current is passed through the wire, the compass needle is deflected.

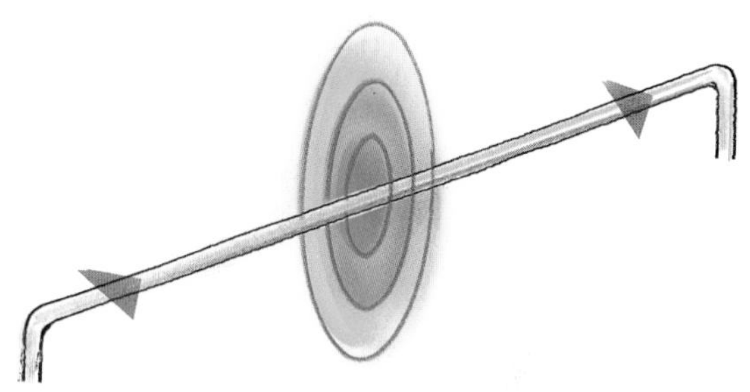

▲ A circular magnetic field exists around a wire carrying a current.

The magnetic connection

The early 1800s was an exciting time for scientists experimenting with electricity. The newly invented battery enabled scientists to have a supply of electricity "on tap" for the first time. Discoveries came thick and fast from scientists around the world – Georg Ohm in Germany, Andre Ampere in France, Michael Faraday in England, Hans Oersted in Denmark, and Joseph Henry in the United States.

It was in 1820 that Oersted demonstrated the intimate connection between electricity and magnetism. One day he was performing an electrical experiment, and there happened to be a compass near the wire carrying the current. He noticed that when he switched on the current, the compass needle was deflected away from north. He realized that the passage of electric current through the wire was producing magnetism.

Oersted's observation triggered off a new branch of science – electromagnetism. This soon led to the invention of the electric generator and electric motor, the key devices on which a vast electrical industry grew up (see Chapter 3).

INVESTIGATE

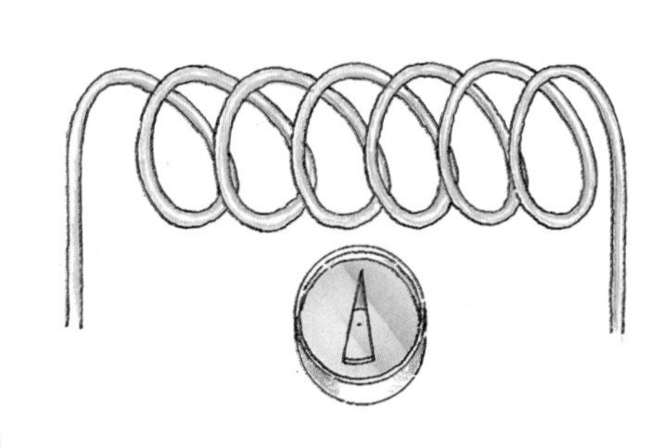

What happens to a compass needle when you place it near a coil of wire carrying an electric current? Find out for yourself.

Make a coil of wire, and place it next to a compass, as shown. (Quick: In which direction is the axis of the coil now pointing?) Now join the ends of the wire to a battery. What happens to the needle of the compass?

Now place a piece of cardboard over the coil carrying the current, and carry out the experiment with iron filings you did with a magnet (see page 16). What results do you get?

Electromagnets

The magnetism set up by the passage of electric current through a wire is intensified if you twist the wire into a coil. If you then plot the magnetic field around the coil with iron filings or a compass, you find that the coil behaves like a bar magnet. It has north and south magnetic poles (see page 18).

If you now place an iron bar inside the coil, its magnetism increases dramatically. This is the principle behind the electromagnet, used at scrapyards, for example, to pick up cars and other steel scrap. The U.S. scientist Joseph Henry was making powerful electromagnets by the late 1820s. One could lift over 3,000 pounds (1,400 kg). Electromagnets are, temporary magnets. They lose their magnetism when the electric current through their coils is switched off.

Smaller electromagnets are found everywhere. In the home they are found most commonly in the doorbell and in the receiver of the telephone. In the doorbell, electromagnets pull the striker against the bell when the bell button is pressed. In the telephone receiver, the electromagnet's varying magnetism causes a diaphragm to vibrate and give out sound waves.

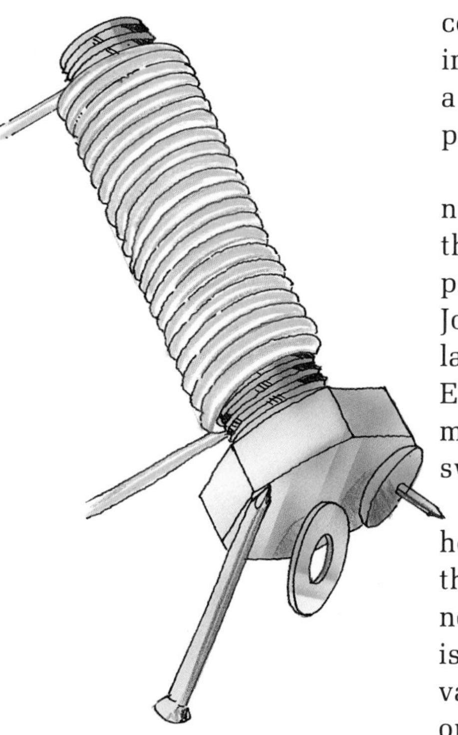

▲ Make yourself a useful electromagnet by winding many turns of insulated copper wire around a large iron bolt. The more turns you have, the more powerful your magnet will be. Attach the ends of the wire to a battery, and see what you can pick up.

Q Why do you need to use insulated wire? Does your electromagnet stay magnetic when you remove the battery?

► This experimental train carries a very powerful electromagnet. It sets up magnetism in a metal rail on the track. The magnetism of the train repels the magnetism of the rail, raising the train above the track. This type of train is known as maglev, which is short for magnetic levitation.

Electromagnetic waves

We saw in the last chapter that an electrically charged object is surrounded by an invisible electric field, a region in which electric forces of attraction and repulsion act. We also saw that a magnet is surrounded by an invisible magnetic field, a region in which magnetic forces of attraction and repulsion act.

But electric and magnetic fields also exist in quite a different form, and in one that we can sometimes see and feel. They exist in the form of waves.

There is a whole family of these electromagnetic waves, often called the electromagnetic spectrum. The electromagnetic waves we can see are light waves; the ones we can feel are heat waves. The Sun puts out these waves, or rays, when it gives off its energy as radiation.

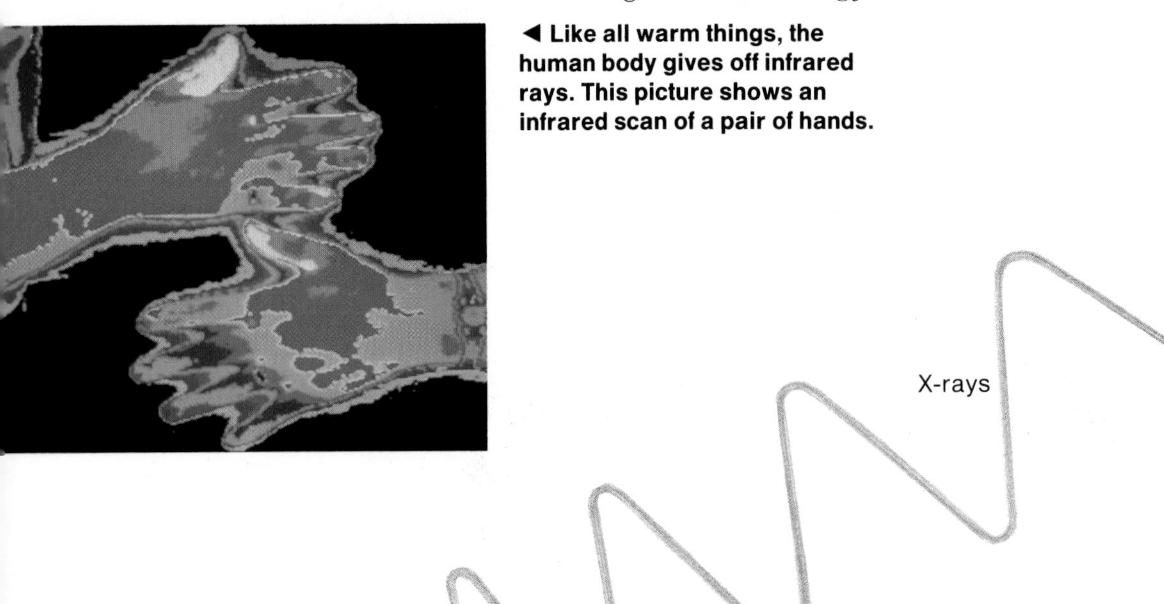

◀ Like all warm things, the human body gives off infrared rays. This picture shows an infrared scan of a pair of hands.

▲ This dramatic picture of San Francisco Bay was taken by an aircraft flying at 50,000 feet (17 km). It is a composite picture produced from photographs taken at different wavelengths through different colored filters. It is printed in artificial colors.

► A radio antenna, which transmits a range of short and medium radio waves. These waves can travel a very long way, being reflected over the horizon by electrified layers in the upper atmosphere.

The electromagnetic family

Light rays and heat rays differ from each other in wavelength, or the distance between the crest (top) of one wave and the next. Light rays have a shorter wavelength than heat rays. There are other waves with a shorter wavelength than light rays, and there are waves with a longer wavelength than heat rays. All electromagnetic waves travel at the same speed.

As shown in the main illustration, from the shortest to the longest wavelengths, the main waves in the electromagnetic spectrum are: cosmic rays, gamma rays, X-rays, ultraviolet rays, infrared (heat) rays, microwaves, and radio waves. The shortest rays have a wavelength of less than one-million-millionth of a meter. The longest radio waves have a wavelength of several kilometers. (Wavelengths are always expressed in metric measurements.)

Q If light travels at a speed of 186,000 miles (300,000 km) a second, how long does it take light from the Sun to reach the Earth, 93 million miles (150 million km) away?

3 Electricity at Work

◀ **Inside the Stanford Linear Accelerator (SLAC) in California. This powerful atom-smashing machine consumes vast amounts of electricity in accelerating atomic particles along a 2-mile (3-km) long underground tunnel.**

By exploiting the connection between electricity and magnetism, we can build powerful machines that can generate a steady supplv of electricity for use in our homes and factories.

We put this electricity to work in thousands of different ways. We use it for lighting and heating, and for powering all kinds of household appliances, from electric toothbrushes and food mixers to refrigerators and vacuum cleaners. Most of these appliances include machinery powered by electric motors.

Both household current and batteries are used to power the numerous electronic devices now found in our homes, such as radio, television, VCRs, computer games, CD players, and so forth. Many of these devices include, and could not exist without, microchips. These thin crystal wafers, about the size of a thumbnail, are miracles of the modern age.

This chapter gives an overview of our electrical and electronic world, but ends with a bang, or rather a mighty collision between particles accelerated by the most powerful electrical machines that exist – atom-smashers.

▼ **An electric hair drier is one of the many useful electrical gadgets found in the home. It uses electricity in two ways. Electricity powers a small motor to turn the fan. The fan draws in cool air and blows it past a series of heating coils. Electricity flows through the coils and heats them up.**

Q The coils heat up because of their high electrical ______. (Fill in the missing word.)

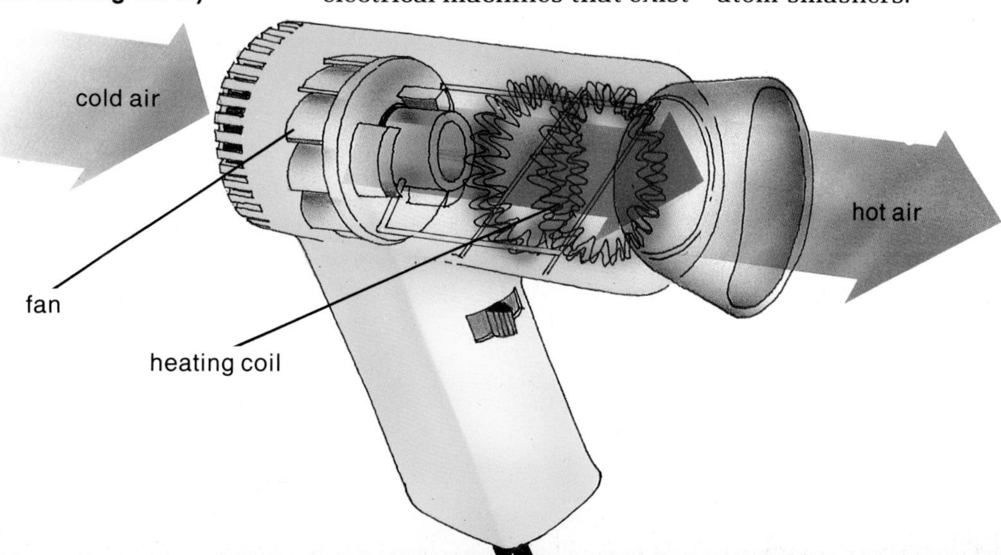

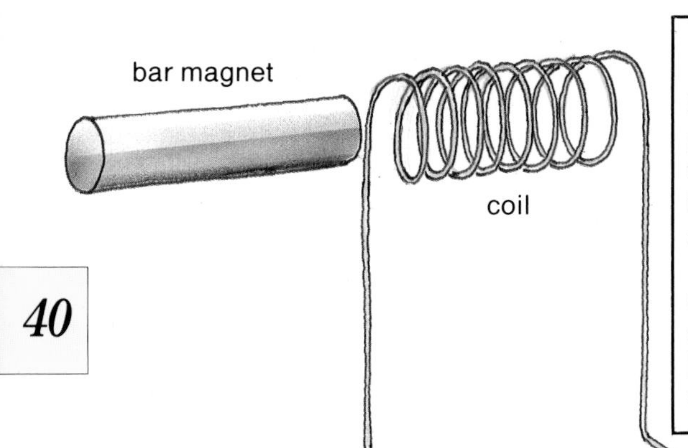

INVESTIGATE

For this investigation you will need a bar magnet, a compass, and copper wire. Wind the wire around the compass to make a current detector. Then make a separate coil, as shown. Set up the compass with the needle pointing north. Don't use the magnet yet.

Now take the magnet and push it into the coil. Look at the compass while you are doing this. Is there any change? Pull out the magnet from the coil, again looking at the compass. Does anything happen? Move the magnet in and out of the coil, first slowly, then quickly. What happens to the compass now?

copper wire

Generating electricity

In 1831, the British scientist Michael Faraday carried out an experiment much like the one above. He showed that when a magnet is moved in relation to a coil of wire, electric current is induced (set up) in the coil.

This phenomenon, known as electromagnetic induction, is the principle behind the electricity generator. In practical generators, coils of wires are moved in a stationary magnetic field, instead of the other way around.

The electricity generators in power stations, which produce the electricity for our homes, have their coils spun by powerful turbines. They are known as altrenators because they produce alternating electric current (AC). This is current that surges first in one direction and then the other. The diagrams on the opposite page show the principle of the alternator and how the current reverses as the coil turns.

Generators can also be made to produce one-way, or direct current (DC). They are fitted not with slip rings, but with a split-ring device called a commutator. DC generators are often called dynamos. Small dynamos can be fitted to bicycles to run the lights. They are usually spun by a small wheel that presses against the side of one of the tires.

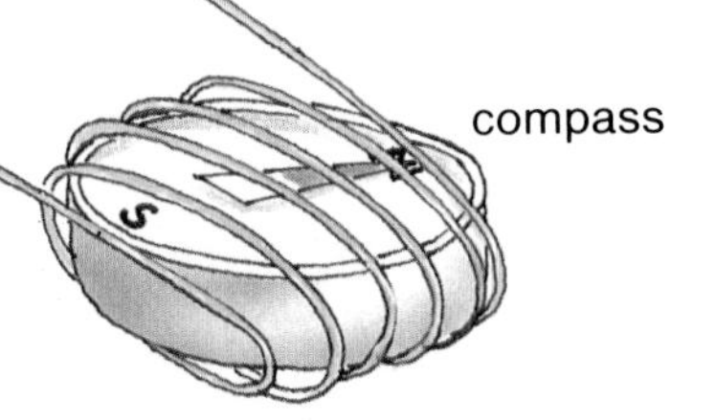

▲ This current detector is made by winding coils of wire around a compass. When current flows in the coil, a magnetic field is set up. This field moves the magnetized needle of the compass.

Q The scientific name for a current detector is galvanometer. Who is it named after?

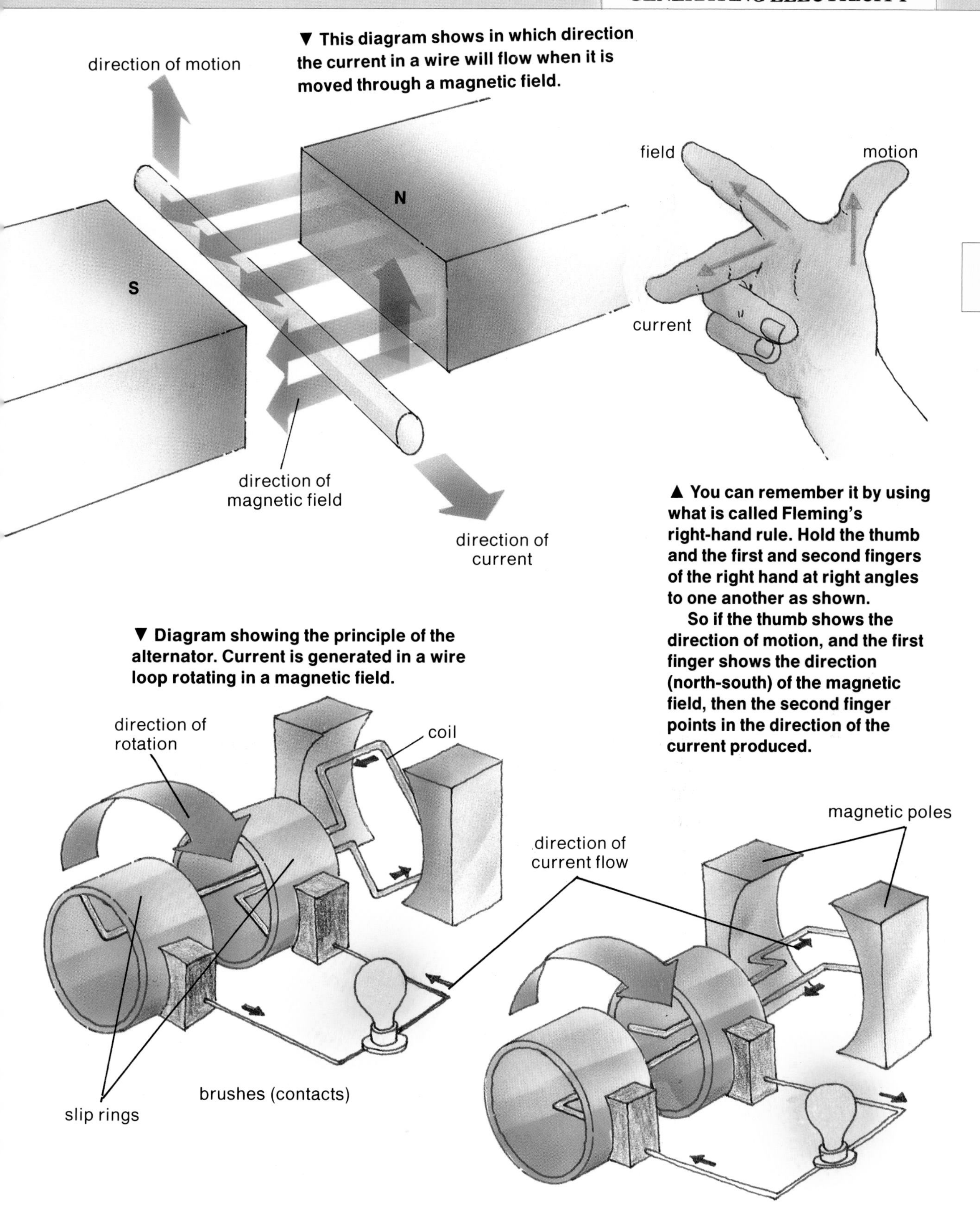

▼ This diagram shows in which direction the current in a wire will flow when it is moved through a magnetic field.

▲ You can remember it by using what is called Fleming's right-hand rule. Hold the thumb and the first and second fingers of the right hand at right angles to one another as shown.

So if the thumb shows the direction of motion, and the first finger shows the direction (north-south) of the magnetic field, then the second finger points in the direction of the current produced.

▼ Diagram showing the principle of the alternator. Current is generated in a wire loop rotating in a magnetic field.

Three main energy sources are used to provide the power to spin the electricity generators at power stations. They are fossil fuels, particularly coal; flowing water in hydroelectric programs, and nuclear energy.

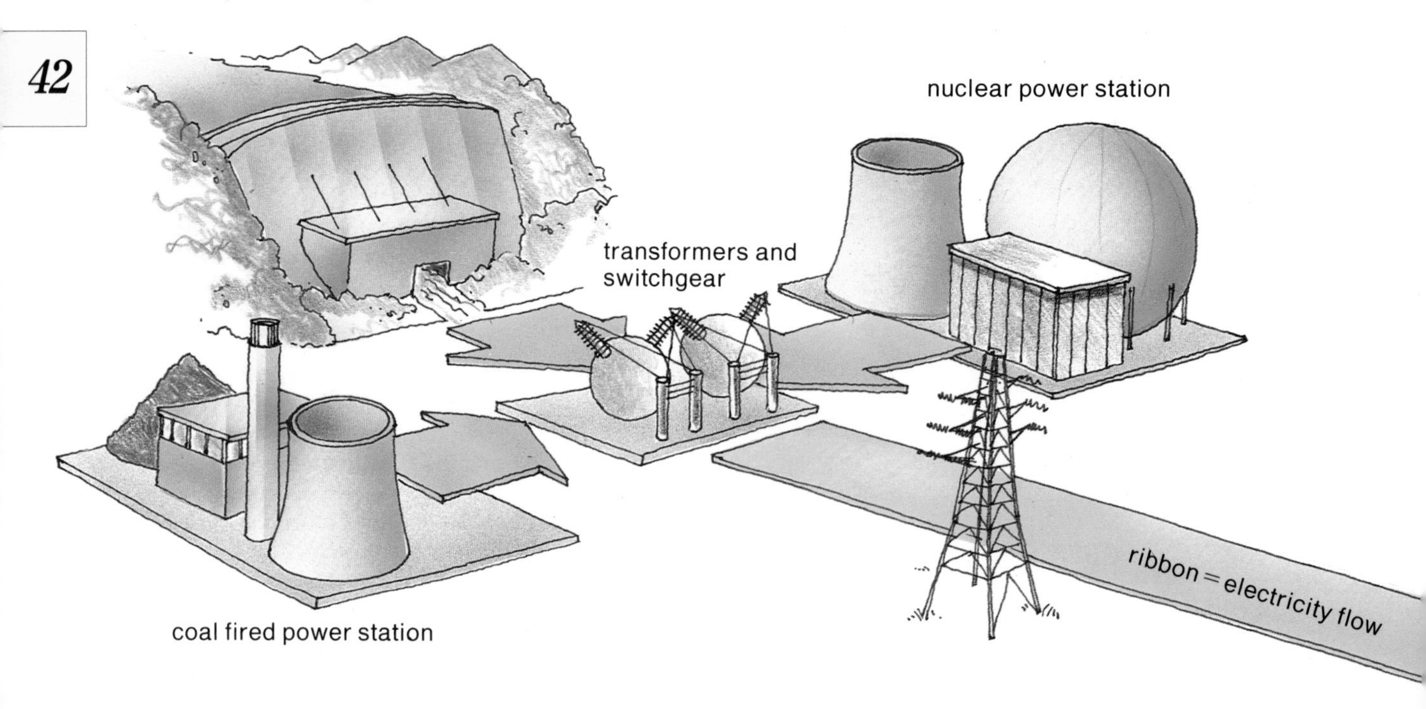

Power to the people

Electricity is produced by power stations dotted around the country. It travels to the factories, businesses, and homes that use it mostly by overhead power (transmission) lines. Often a number of power stations are linked together into a grid system serving a large region. This makes it possible to use electricity produced in one area to meet demand in another.

Power stations produce alternating-current (AC) electricity at a voltage of about 25,000 volts. If the electricity were transmitted along the power lines at this voltage, a great deal of power would be lost because of heating effects. But, strangely enough, if the voltage is greatly increased, the heating effect and energy loss are greatly reduced.

Step up, step down

Transmission lines may carry electricity at voltages of 500,000 volts or more to avoid energy loss over long distances. The generated voltage is increased to this level by means of transformers. Transformers are also used at substations to reduce the voltage to the levels required by users.

The transformer is a most useful electromagnetic device. It consists of two coils, wound around an iron core. One coil has more turns that the other. By feeding AC at a certain voltage to the coil with fewer turns, AC comes out of the coil with more turns at a higher voltage. This type is called a step-up transformer. On the other hand, when you feed AC to the coil with more turns, AC comes out of the other coil, with fewer turns at a lower voltage. This type is called a step-down transformer.

Electricity is transmitted at high voltage overland. It is reduced in voltage at various substations, which feed the lower voltages to cities, towns, and industries. Industries use the highest-voltage electricity – up to about 35,000 volts. Homes require, typically, only 120 or 240 volts.

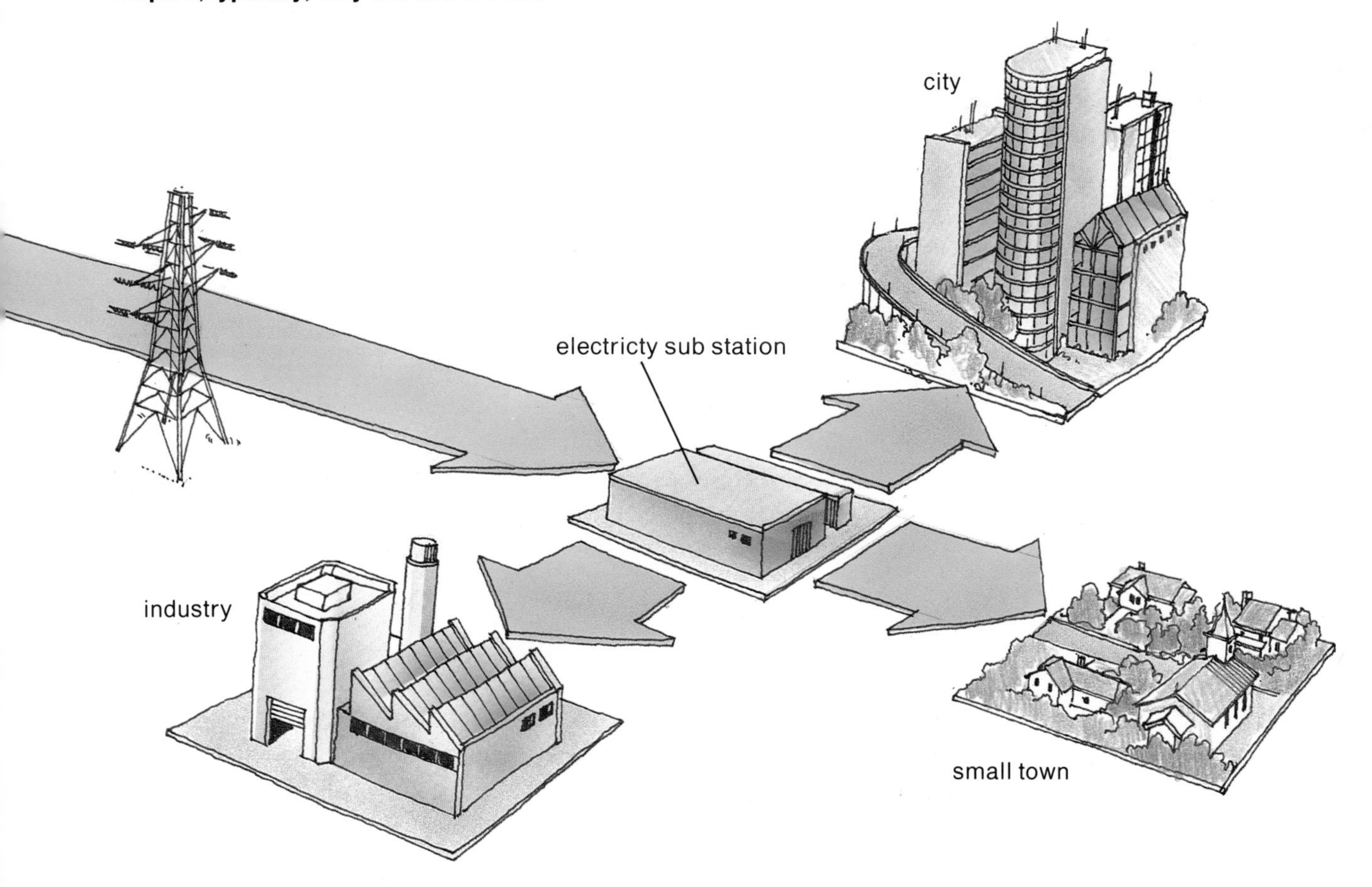

► **This diagram shows in which direction a wire will move when it is carrying a current in a magnetic field. You can remember it by using what is called Fleming's left-hand rule.**

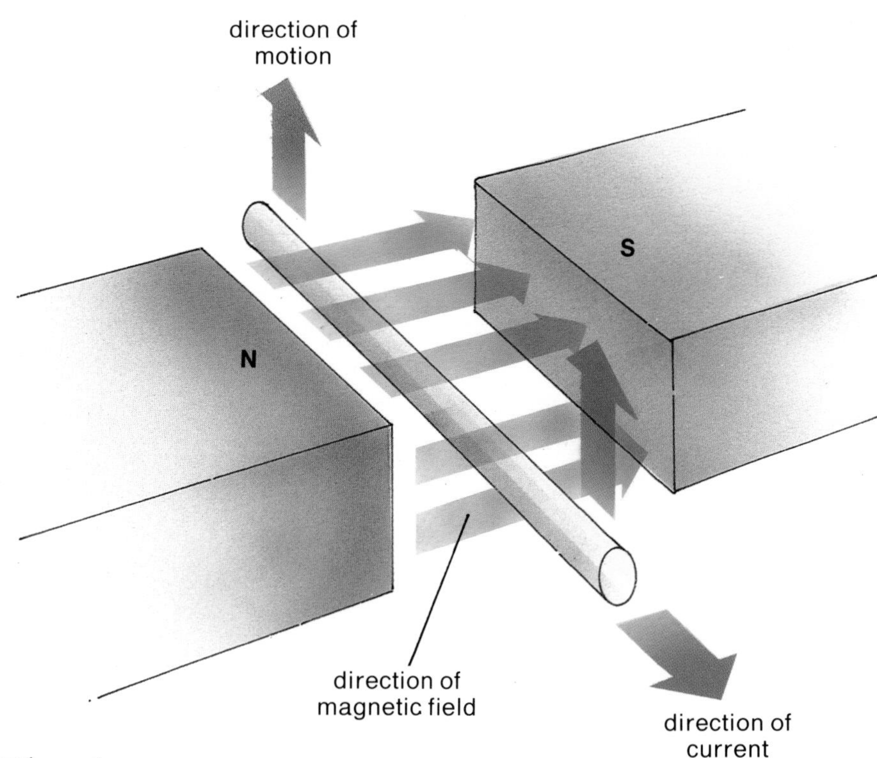

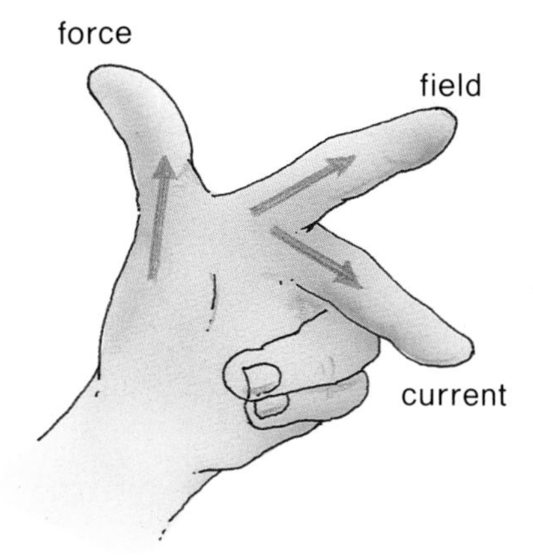

Hold the thumb and the first and second fingers of the left hand at right angles to one another as shown.

So if the first finger shows the direction of the magnetic field, and the second finger points in the direction of the current, then the thumb shows the direction of motion.

Electric motors

We have seen how electric current is produced in a wire when it is moved through a magnetic field (see page 41). The opposite effect also occurs. When you pass electricity through a wire that is in a magnetic field, the wire moves. This is often called the motor effect, and it is the principle behind the electric motor.

Electric motors power many of the machines we use in the home, from electric toothbrushes, food mixers, and model electric trains, to drills, washing machines, and hedge trimmers. Electric motors also drive industrial machinery such as lathes, some road vehicles, and many locomotives.

Motor design

Electric motors are built in much the same way as electric generators. They consist of many coils of wire wound on a shaft (rotor), which is free to rotate between the poles of a magnet. Electric current is passed through the coils. The magnetism set up by this current interacts with the surrounding magnetic field and pushes and pulls the coils, and this makes the rotor rotate.

▲ **Electric motors power the "bullet trains" that run on Japan's high-speed railroad network, called the Shinkansen. Power cars in the train pick up electricity from overhead power lines at a voltage of 25,000 volts AC.**

Q Can you remember what AC stands for?

As each coil on the rotor rotates, it travels first in one direction, then in the opposite direction, in relation to the surrounding magnetic field.

To keep it moving the same way, the current through it must be reversed every time it changes direction. In motors worked from the direct (one-way) current from batteries, a split-ring device called a commutator reverses the current (see Box).

AC motors

Most electric motors, however, work from alternating (two-way) current. The most common type is the induction motor. This type has a rotor made up of conducting bars in the shape of a cylinder (called a squirrel cage).

Alternating current is fed not to the rotor, but to surrounding coils. These coils become magnetic, and this induces current, and in turn magnetism, in the rotor. The magnetic fields of the coil and the rotor interact, and the rotor is pushed and pulled around and around.

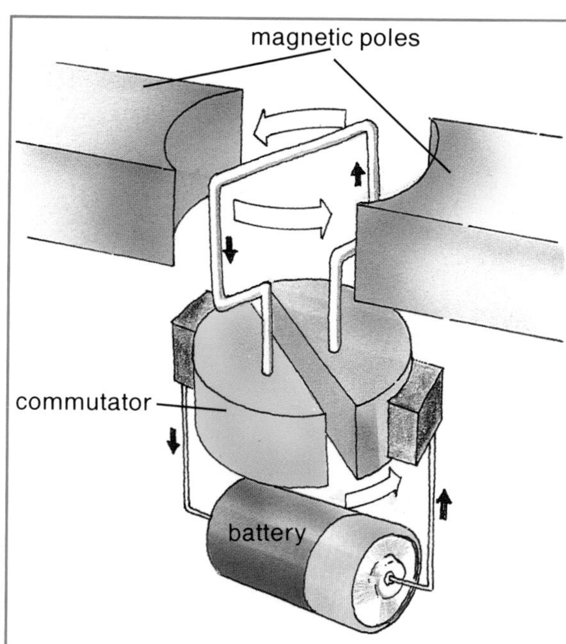

The commutator

Electric motors that work from the direct current (DC) from batteries are fitted with a commutator. This device reverses the direction of the current in the coil every half-turn. This ensures that the coil will always experience forces that keep it rotating in the same direction.

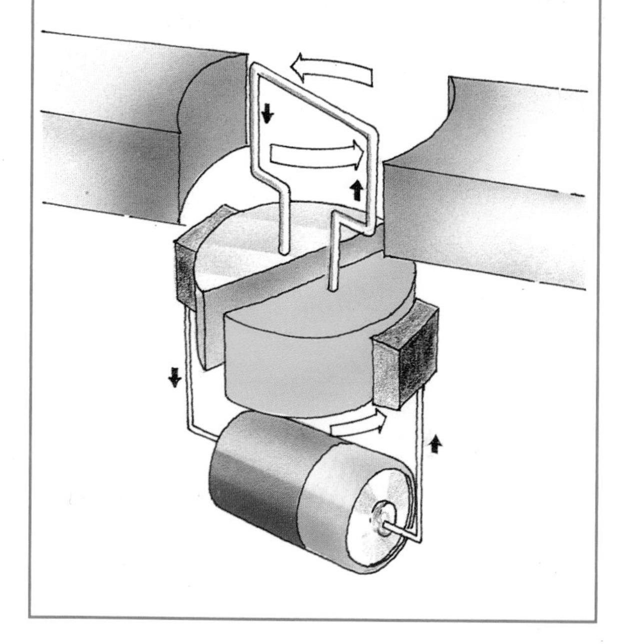

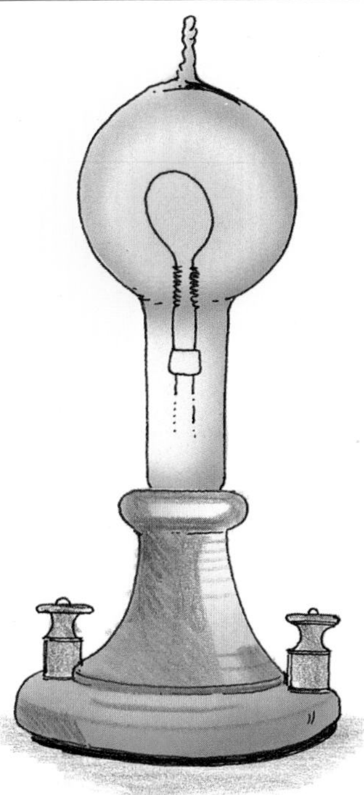

▲ **An early Edison electric lamp. It used baked thread as a light-producing filament.**

► **A modern light bulb, which has a coiled filament and is gas filled.**

Electric lights

The electric light, perhaps one of the greatest human inventions, allows us to prolong the day beyond sunset for work and leisure.

The oldest form of electric light, the light bulb, was first demonstrated by Joseph Swan in Britain in 1878 and by Thomas Edison in the U.S. a year later. In 1883, the two men joined forces to mass-produce bulbs.

The light bulb is technically called the incandescent filament lamp. "Incandescent" means something like "glowing because of heat." The lamp consists of a glass bulb containing a thin wire, or filament. The filament has a high electrical resistance, and when electricity is passed through it, it heats up and produces light.

The filament is made of tungsten, the metal with the highest melting point (6,116°F, 3,380°C). The bulb is filled with a mixture of nitrogen and argon gases.

Q Why would it not be a good idea to fill light bulbs with air?

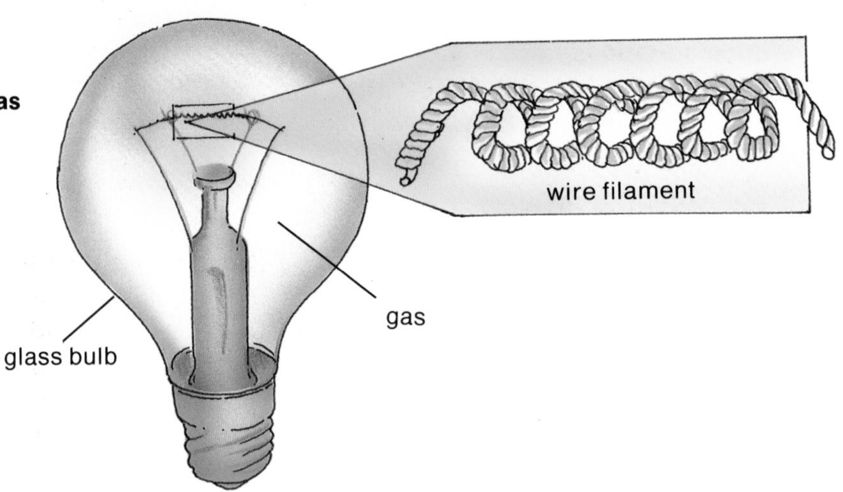

▼ **The essential features of a fluorescent tube, which contains mercury vapor at low pressure and has a fluorescent coating on the inside.**

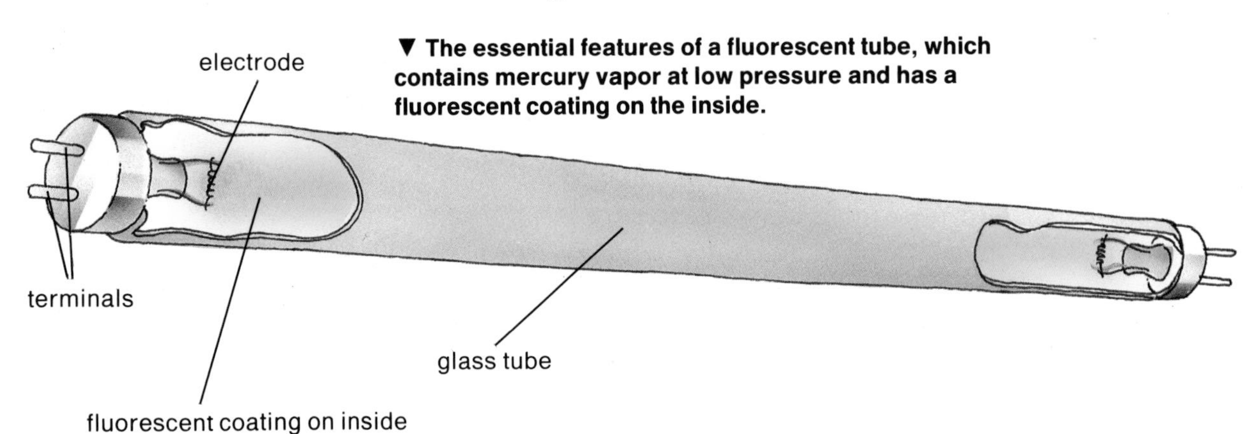

► **Neon and other discharge tubes are combined to create this colorful hotel sign. Such tubes can be fashioned into any shape, which makes them ideal for advertising purposes. By using different gases inside and different coatings outside, different colors can be produced.**

Discharge lamps

Other kinds of electric lights produce light in quite a different way. They consist of a tube filled with gas, which gives out light when electricity is passed through it.

Ordinarily, gases are not good conductors of electricity. But they become good conductors when they are at very low pressure. When electricity is passed through a gas at very low pressure, an electric discharge takes place. The atoms of the gas become "excited" and give off energy in the form of light.

If the gas is neon, the light produced is a brilliant orange-red. This is a favorite light for advertising signs. If the gas is sodium vapor, the light produced is a brilliant yellow.

If the gas is mercury vapor, the light produced is invisible ultraviolet light. This vapor is used in the fluorescent lamp. The tube that forms the lamp has a fluorescent coating on the inside. This gives out visible white light when the ultraviolet light hits it.

Our electronic world

Electronics is one of the most recent and most exciting branches of electrical science. Radio, television, video cassette recorders, Walkmans, CD players, fax machines, digital watches, and pocket calculators are some of the many electronic devices we now use.

But exactly what is "electronics," and how does it differ from ordinary "electrics"? Both involve the flow of electric current, that is, of electrons. But "electrics" generally involves the steady flow of electrons through wires. In "electronics," the flow of electrons is manipulated in all kinds of ways by different devices. In these devices, the electrons may flow through a vacuum, through gas, or through crystals.

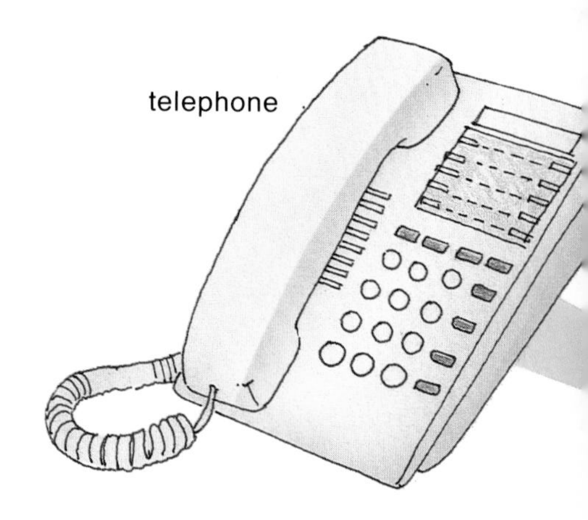

Not so long ago the only electronic device in the home was the radio. Today, many homes have a huge variety of electronic equipment, such as those illustrated here. Almost all these devices now rely on microchips for their operation.

The electron tube

The first devices used to manipulate electrons were electron tubes. They contained metal electrodes sealed in a glass tube from which the air had been removed.The flow of electrons from one of the electrodes, when it was heated, could be controlled by altering the electric charge on the other electrodes.

Electron tubes were pioneered by Ambrose Fleming in Britain and Lee de Forest in the U.S. in the early years of this century. Their use led directly to the development of radio broadcasting.

Q Electron tubes were also called vacuum tubes. Why was this a good name for them?

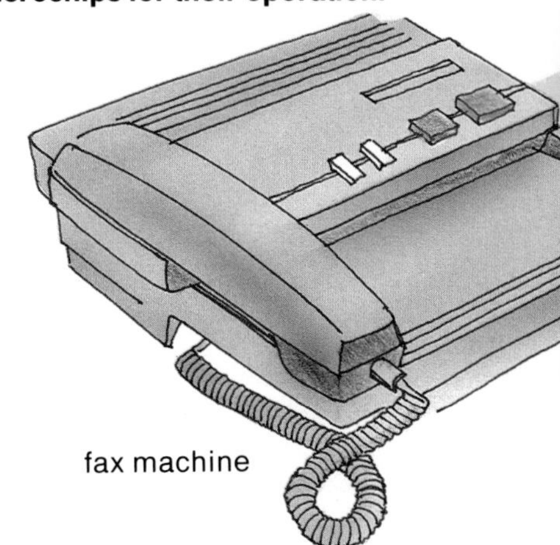

The transistor revolution

In 1948 came the invention of a device that made the electron tube obsolete almost overnight. This device was the transistor, developed by the U.S. team of John Bardeen, Walter Brattain, and William Shockley. It won for them the 1956 Nobel physics prize.

The transistor was pill-sized, a fraction of the size of a typical electron tube. It consisted of three layers of semiconductor crystal sandwiched together. Later, transistors and other electronic components were incorporated on a single crystal wafer. This led soon to today's miraculous electronic component, the microchip.

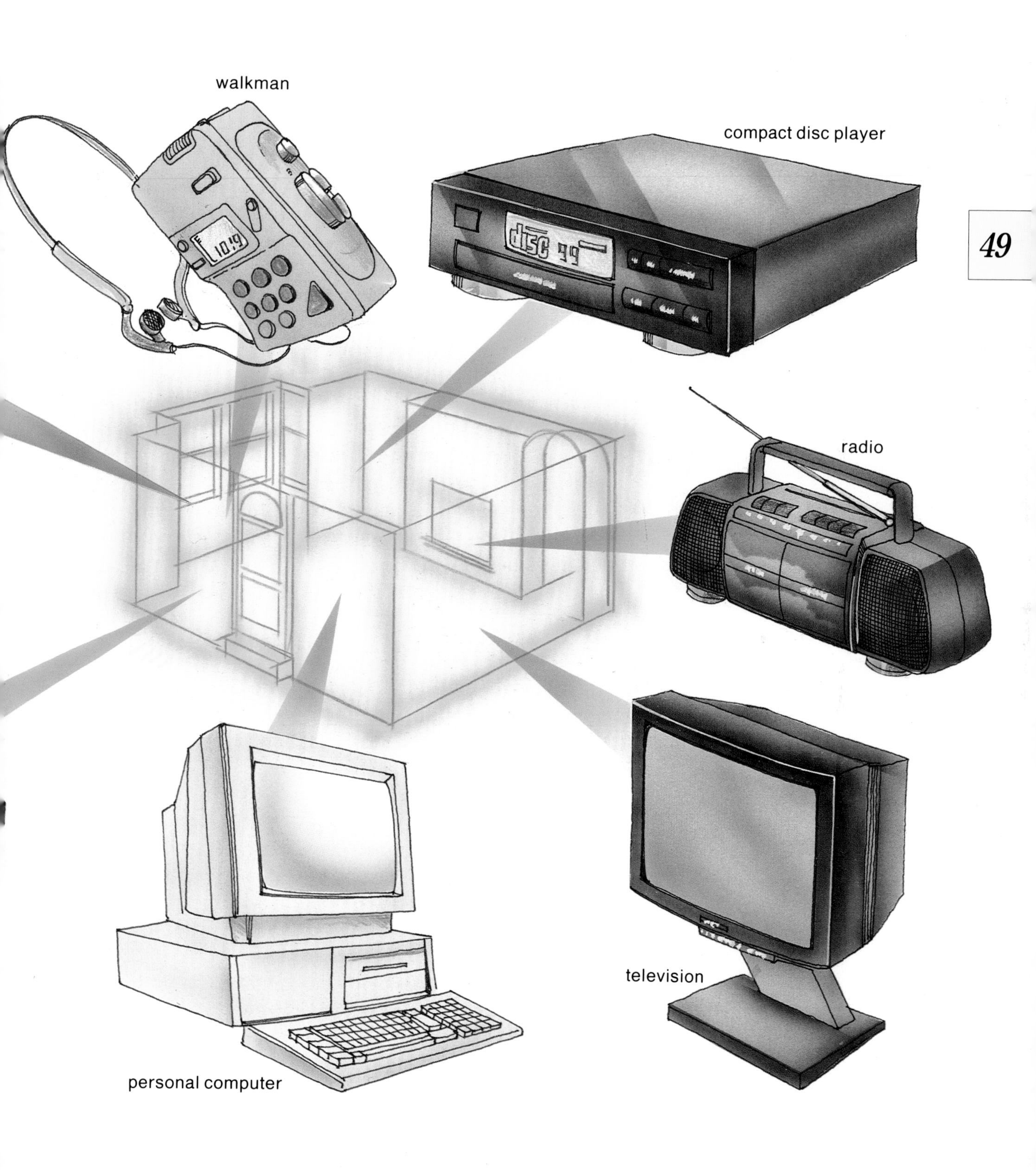
walkman
compact disc player
radio
television
personal computer

The microchip

Tiny wafers of crystal smaller than a fingernail have brought about the revolution that is still taking place in electronics. These tiny wafers or chips of silicon are the most common kind of semiconductor.

In particular, silicon chips have made possible compact but powerful computers. Many of the personal computers people now have at home are more powerful than the giant computers large business corporations used only about 20 years ago!

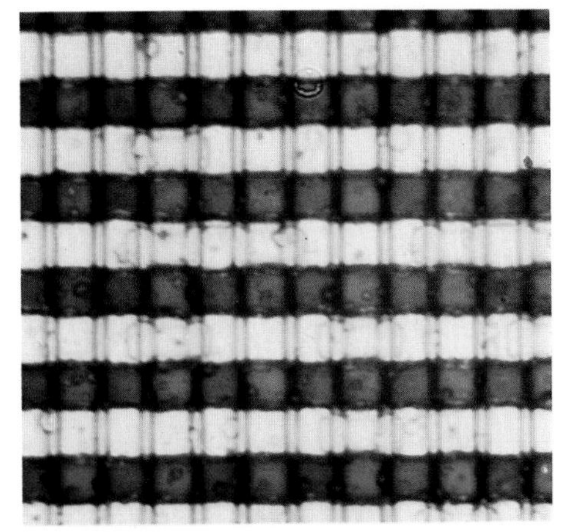

▲ A highly magnified picture showing some of the electronic circuits on a silicon chip. The circuits and the circuit components are all integrated in the same thin crystal wafer.

Computers and other complicated electronic devices need many electronic circuits to work. Silicon chips are so useful because they each contain thousands of electronic circuits. These circuits are formed within the silicon itself and are known as integrated circuits. They are made up of devices such as transistors, capacitors, and resistors, and also the connecting paths between them.

Each component is microscopic in size, which is why the silicon chip is often called a microchip. Some of the latest microchips contain up to a million separate components.

Making the micro

The individual components in a microchip are made up of a number of layers of silicon. Each layer has been given different electrical properties by treatment with chemicals. The treatment process is called doping.

In chipmaking, chips are made, hundreds at a time, on a thin disk of silicon about 4 inches (10 cm) across. The circuits are built up, layer by layer, in a lengthy series of steps.

In each layer, the silicon needs to be doped or otherwise processed only in certain areas. So doping takes place following a masking process, which allows only the desired areas to be affected.

When all the layers are completed, metallic connections are made between the components by depositing a thin film of aluminum or gold. After being tested by automatic probes, the silicon disk is cut up into individual chips. They are mounted on a plastic base, and fine gold wires are used to connect the chip circuits to pins on the base. The chip units are now ready for use.

Essential steps in chipmaking

1.

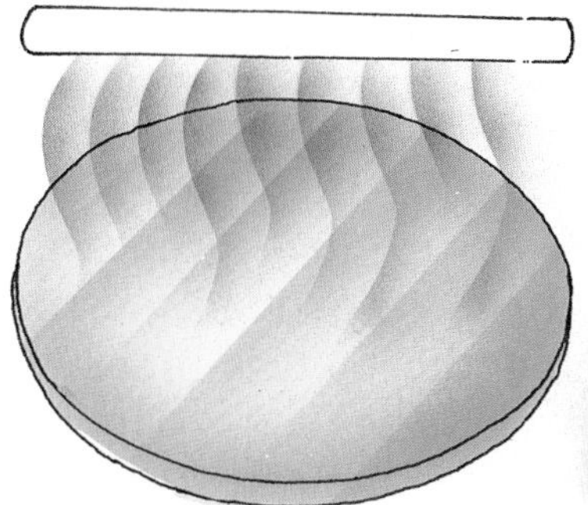

▲ **1. Doping**
At the start of the chipmaking process, the pure silicon disk is doped or treated, with a chemical called boron. Later, areas of the silicon will be treated with phosphorus.

2.

◀ **2. Insulating**
Heating the disk in a very hot furnace produces a layer of silicon dioxide. This acts as insulation between the various components. The disk goes through this process several times.

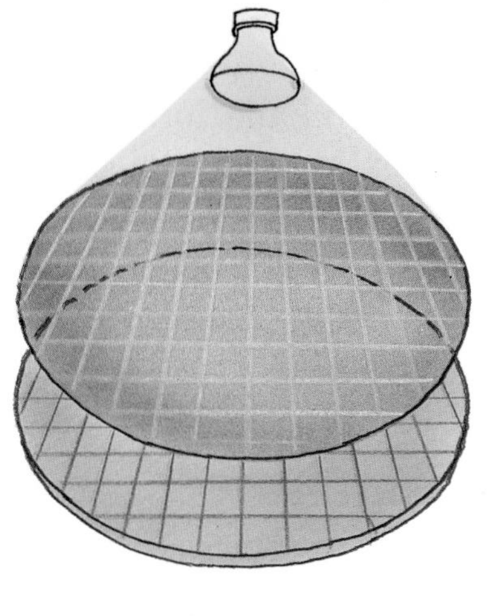

3.

► **3. Masking**
The masking processs consists of several stages. The disk is coated with a light-sensitive layer, then light is shone onto it through a mask. The light hardens exposed areas, but unexposed areas remain soft and can be washed away.

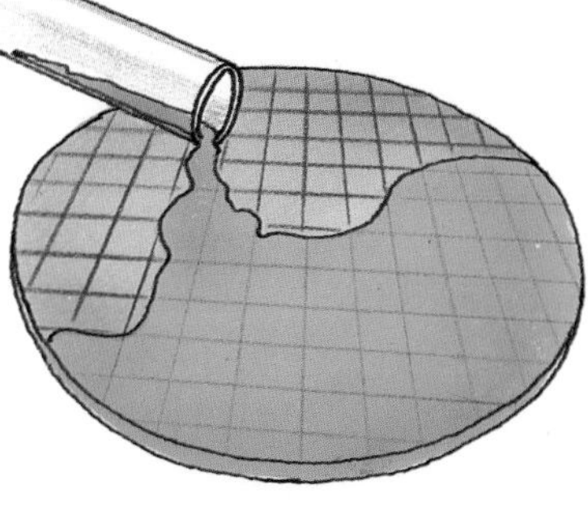

4.

▼ **4. Etching**
Acid is used to eat away layers exposed by the masking process. The exposed areas are then ready for doping.

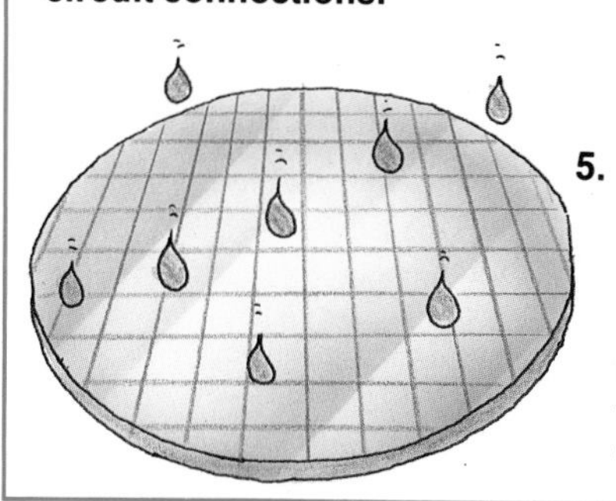

5.

▼ **5. Metal coating**
This is the final stage of chipmaking. An aluminum film is deposited on the silicon disk. It is then masked and etched to provide the necessary circuit connections.

6.

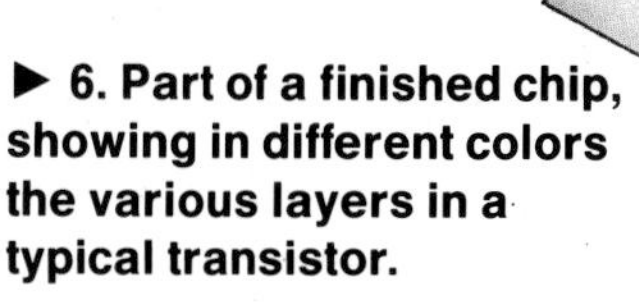

► **6. Part of a finished chip, showing in different colors the various layers in a typical transistor.**

Computer basics

Some of the most powerful computers are used in weather forecasting. They are fed with millions of weather instrument readings from around the world – of temperature, air pressure, wind speed, humidity, and so on. They then process all these figures and calculate how the weather is likely to change in the future. Such computers carry out billions of operations every second, a task that would otherwise require a whole army of meteorologists.

Computers perform phenomenal feats of calculation, but they do not do it in a complicated way. In fact, they work by doing very simple operations, such as addition and subtraction. They achieve their incredible computing power because they carry out these operations at very high speed.

Binary digits

Computers handle data, or information, in the form of numbers, or digits, which is why they are called digital computers. They do not work with the ten ordinary decimal digits, from 0-9. Instead, they work with the two binary digits, 0 and 1. In the computer data is represented as a sequence of 1s and 0s.

In electrical circuits, these two digits can be represented, for example, by flow (1) or non-flow (0) of electricity. This can be achieved by switching electric current on and off. The 1s and 0s can also be represented by an electricity charged (1) or uncharged (0) state.

Simply stated, computers consist of thousands of switches (transistors) and electronically charged or uncharged components (capacitors).

IT'S AMAZING!

During the lift-off of the space shuttle (right), 300,000 items of data have to be processed every second to ensure that the launch proceeds smoothly. That is why the launch is under computer control. Only computers are able to handle such vast amounts of data so quickly. In fact, the shuttle orbiter carries five computers, each of which processes all the data fed to them.

► Students working at a computer terminal. They are using it to extract data from an information store, or database, in another computer. Using different programs they can also use the terminal for a variety of other purposes, for example, word-processing and desktop publishing.

Programs

The computer program, or set of instructions for operating a computer, is written as a lengthy sequence of very simple steps. It is organized so that there are two possible outcomes from each step. These can then be represented by a 1 or a 0, and thus by the electronic circuits in the computer.

Computers only work on data in the form of binary code. But fortunately programmers do not have to write their programs in binary. Instead they write it in a computer language, which the computer recognizes and converts into binary. There are several computer languages, including BASIC for ordinary use, COBOL for business use, and FORTRAN for scientific use.

Essential computer terms

BITS – Binary digits. The 1s and 0s of the binary code, the number system used in computing.

DATA – Information in the form of figures or words fed into the computer.

HARDWARE – The physical units that make up the computer – keyboard, central processing unit, visual display unit (VDU), disk drive, printer, mouse, and so forth.

LANGUAGE – A simplified language computer programmers use to write programs.

MEMORY – Part of the computer in which data is stored. The basic operating program for the computer is stored in a read-only memory (ROM), which can be read, but can't be altered. Instructions and data for processing are stored in the random-access memory (RAM), which can not only be read, but also be added to, changed, or deleted.

PROGRAM – The set of instructions that enables the computer to operate.

SOFTWARE – The information and instructions fed into the computer.

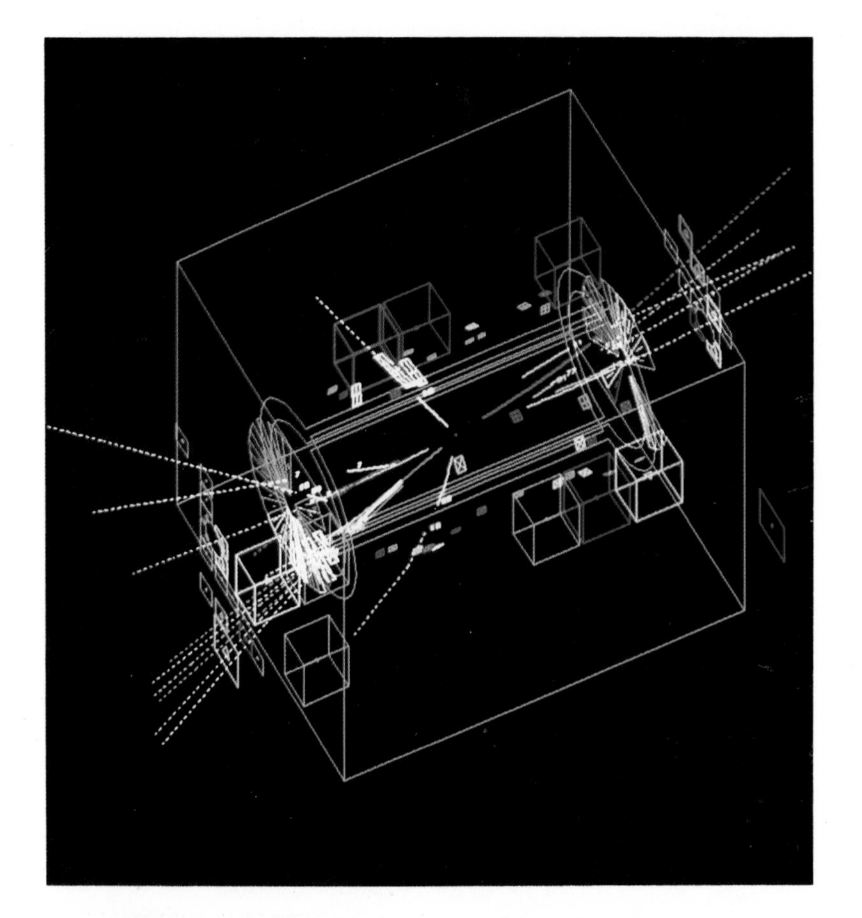

► **Computer graphics have become an important tool in science and engineering. This "3-D" image shows the paths of atomic particles through a detector during a nuclear physics experiment.**

Smashing atoms

The most powerful electrical machines of all are used by scientists to probe into the nature of matter. They use these machines literally to smash atoms and atomic particles to pieces. That is why these machines are popularly called atom-smashers.

But the correct name for these machines is particle accelerators, because that is what they do – accelerate atomic particles. They accelerate the particles and then either direct them at a target so that they penetrate its atoms, or collide the particles with another accelerated particle beam. In each case a variety of new particles is produced.

Particle accelerators use charged particles as "ammunition" for their atomic bombardment. They mostly use protons (positive charge) and electrons (negative).

Q Why don't they use neutrons, the other main particle found in atoms?

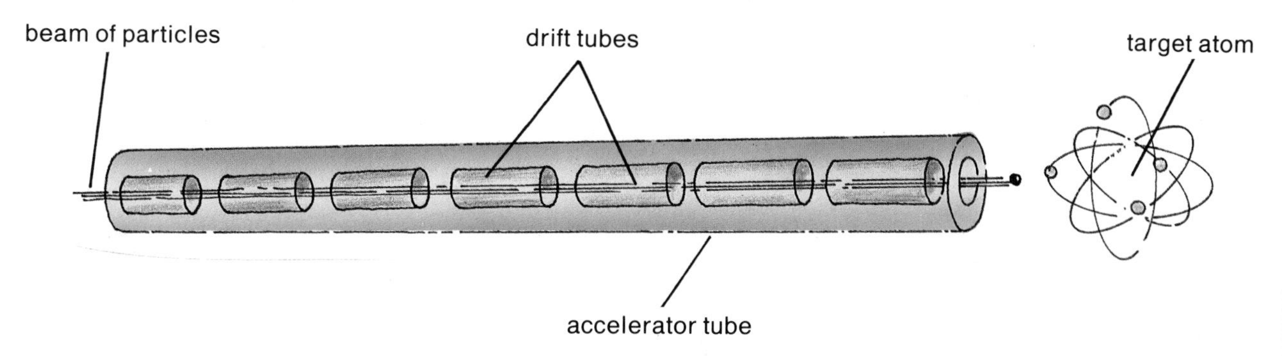

Particle race tracks

There are two main kinds of particle accelerators – linear and circular. In a linear accelerator, particles are accelerated in a straight line. Powerful electric fields are applied all the way along, which alternately pull and push the particles and so speed them up. This uses the simple principle that electric charges can attract or repel one another (see page 12). The U.S. has one of the world's most powerful linear accelerators, Stanford University's linear collider (SLAC) in California.

Particle beams can be given more energy if they are made to travel around and around in a circle and accelerated every time they go round. The particles are made to travel in a circle by means of a ring of powerful electromagnets. One of the most powerful circular accelerators is the Tevatron, which is operated by Fermilab, near Chicago.

▲ The principle of a linear accelerator. Charged particles are accelerated along a tube by powerful electric fields. The fields alternately pull then push the particles, until they are traveling at very high speed.

▼ In a circular accelerator, charged particles are accelerated by powerful electrical fields. They are kept traveling in a circle by the magnetism of rings of powerful electromagnets.

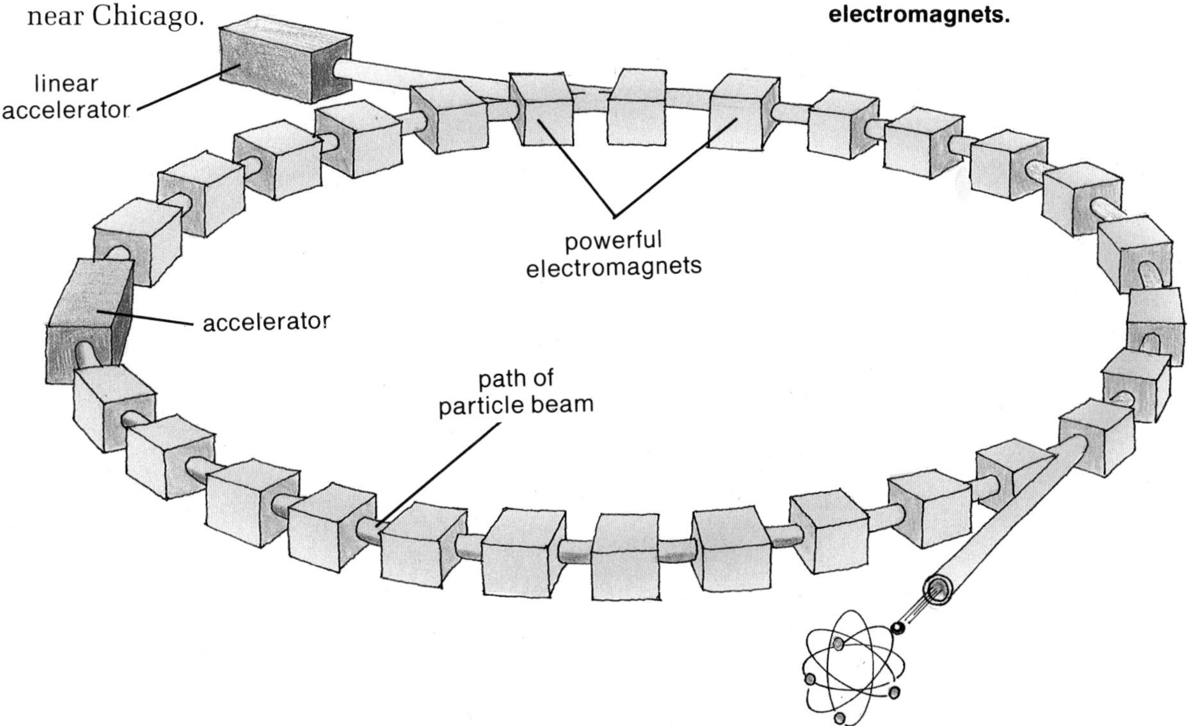

ABOUT 1100 Chinese sailors were using pieces of lodestone as a compass for navigating.

1600 William Gilbert in England showed that the Earth behaves like a magnet. He discovered that the magnetism of each end of a magnet was different, introducing the idea of magnetic "poles."

1752 American scientist and statesman Benjamin Franklin showed that lightning is electricity by flying a kite in a thunderstorm.

1785 Charles Coulomb in France investigated electric charges and magnetism.

1786 Luigi Galvani in Italy demonstrated "animal electricity" by experimenting with frogs' legs.

1800 Alessandro Volta invented the voltaic pile, the first electric battery.

1820 Hans Christian Oersted in Denmark discovered the principle of electromagnetism.

1826 Georg Simon Ohm in Germany stated the law of electrical resistance now named after him.

1831 Independently Michael Faraday in Britain and Joseph Henry in the U.S. discovered the principle of electromagnetic induction.

1859 French physicist Gaston Plante invented the lead-acid battery, or accumulator.

1868 French scientist Georges Leclanché developed the cell named after him, which is the forerunner of the modern dry cell.

1886 Heinrich Hertz in Germany proved the existence of electromagnetic waves.

1878 Joseph Swan in Britain demonstrated the first practical electric lamp.

1879 Thomas Edison in the U.S. produced his electric lamp.

1882 The first public electric power stations began operating in London and New York City. Thomas Edison developed the system for New York.

1888 Croatian-born scientist Nikola Tesla, working in the U.S., developed a practical alternating current electric motor.

1897 English physicist J. J. Thomson discovered the electron.

1911 Dutch physicist Heike Onnes discovered superconductivity.

1936 The Boulder (later named the Hoover) Dam was completed across the Colorado River on the Arizona-Nevada border for power generation and flood control. It is still one of the country's leading hydroelectric plants.

1948 John Bardeen, William Shockley, and Walter Brattain in the U.S. develop the transistor.

1951 Electricity was first generated using nuclear energy at Arco, in Idaho.

1954 Bell Telephone Laboratories in the U.S. developed the photovoltaic (solar) cell.

1958 Texas Instruments in the U.S. introduced integrated circuits.

1965 The world's biggest power blackout occurred on November 9-10, plunging 30 million people in seven northeastern states and Ontario, Canada, into darkness.

1971 Intel Corporation and Texas Instruments introduced the microprocessor.

1991 First controlled nuclear fusion took place in a tokamak, or circular chamber, known as JET (Joint European Torus), located near Oxford, England.

Glossary

ALTERNATING CURRENT (AC) Electric current that flows first one way, then the other. It is this kind of electricity that is supplied to our homes. Compare to **DIRECT CURRENT.**

ALTERNATOR An electric generator that produces alternating electric current.

AMMETER An instrument that measures electric current.

AMPERE Or amp; the unit used to measure electric current, named after the electrical pioneer Andre Ampere.

ANODE A positively charged electrode. Compare to CATHODE.

ATOM SMASHER The popular name for a particle accelerator.

BATTERY A device that produces electric current as a result of a chemical reaction.

BITS Binary digits; the 1s and 0s of the binary code, the number system used in computing.

CAPACITOR A device that can store electric charge.

CATHODE A negatively charged electrode. Compare to **ANODE.**

CATHODE RAYS An old term for a stream of electrons.

CHIP A thin slice of semiconductor crystal, usually silicon, containing microscopic electronic circuits.

COMPUTER LANGUAGE A simplified language computer programmers use to write programs.

COMMUTATOR A split-ring device used in electric motors and generators to change the direction of electric current as the motor or generator turns.

CONDENSER Alternative term for **CAPACITOR.**

CONDUCTOR A material that passes on, or conducts, electricity.

CURRENT, ELECTRIC A flow of electrons through a conductor.

CYCLOTRON A particle accelerator in which particles are accelerated in a spiral.

DATA Information in the form of figures or words fed into a computer.

DIRECT CURRENT (DC) Electric current that flows in one direction only. Compare to **ALTERNATING CURRENT.**

DOPING Introducing chemical impurities into a semiconductor; a stage in chip manufacture.

DRY CELL The most common kind of battery, found in flashlights, radios, and the like. Its chemicals are sealed inside a casing, hence it is dry.

DYNAMO An electric generator that produces direct current.

ELECTRIC ARC A discharge of electricity between two electrodes; in effect a continuous electric spark.

ELECTRIC GENERATOR A machine that produces electricity. It works on the principle that a current is set up in a conductor when the conductor is moving in a magnetic field.

ELECTRIC MOTOR A motor powered by electricity. It works in the opposite way to an electric generator – on the principle that a magnetic field exists around the generator when it is carrying an electric current. The force of this magnetic field is used to turn the armature of the motor.

ELECTROCHEMISTRY The branch of

ELECTRODE The terminal in electric apparatus that conducts electricity in or out. See also **ANODE, CATHODE.**

ELECTROLYSIS The splitting up of a chemical compound by passing electricity through it. It is used to prepare many chemicals and is the principle behind electroplating.

ELECTROLYTE A substance that conducts electricity when it is molten or in solution.

ELECTROMAGNET A temporary magnet consisting of a set of coils, wound on an iron core. The device becomes magnetic when electricity is passed through the coils.

ELECTROMAGNETIC INDUCTION The principle on which electric generators and motors work – that the passage of electric current through a conductor sets up a magnetic field.

ELECTROMAGNETIC WAVES Radiation that consists of traveling waves of electric and magnetic disturbances. X-rays, light rays, and radio waves are among the many kinds of electromagnetic waves.

ELECTROMAGNETISM The study of the relationship between electricity and magnetism.

ELECTRON A tiny particle found in all atoms. An electric current consists of a flow of electrons.

ELECTRONICS The study and application of devices that control the flow of electrons, through semiconductor devices, gases, a vacuum, and so forth.

ELECTROPLATING Coating a material with metal by means of electrolysis.

ELECTROSTATICS The study of static electricity.

FIELD The region around a body where a certain influence is felt, such as an electric or magnetic field, and so forth.

FREQUENCY Of a wave: the number of waves that pass a certain point in a certain time.

FUEL CELL A device that produces electricity as a result of the "burning" of fuel gases, such as hydrogen.

GALVANOMETER A sensitive instrument used to detect electric current, named after the electrical pioneer Luigi Galvani.

HARDWARE The physical units that make up a computer – keyboard, main systems unit, visual display unit (VDU), disk drive, printer, mouse, and so forth.

HYDROELECTRICITY Electricity produced by harnessing the energy in flowing water.

INSULATOR A material that does not conduct electricity.

INTEGRATED CIRCUIT A complete electric circuit – connections as well as components – formed in a single semiconductor crystal. Silicon chips contain thousands of integrated circuits.

MAGLEV Abbreviation for magnetic levitation. Maglev trains are raised (levitated) above the track by magnetic repulsion.

MAGNETIC FIELD The region around a magnet in which the magnetic forces act.

MAGNETIC TAPE A tape coated with magnetic particles, used for recording sound and pictures.

MEMORY The part of a computer in which data is stored. The basic operating program for the computer is stored in a read-only memory (ROM), which can be read, but can't be altered. Instructions and data for processing are stored in the random-access memory (RAM), which can not only be read, but also be added to, changed, or deleted.

MICROCHIP Another name for a silicon chip, which is made up of thousands of microscopic electronic circuits.

OHM The unit of electrical resistance, named after the electrical pioneer Georg Ohm.

PARTICLE ACCELERATOR A machine used for accelerating electrically charged atomic particles to high speed. Popularly called the atom smasher.

PHOTOELECTRIC CELL A cell whose electrical properties change when light falls on it. Such cells are used in camera light meters, television camera tubes, and automatic detection devices.

PHOTOVOLTAIC CELL The correct name for a solar cell, a device in which sunlight sets up a voltage.

POLES The ends of a magnet, where its magnetism appears to be concentrated.

PROGRAM A set of instructions that enables a computer to operate.

RECTIFIER A device that changes alternating electric current into direct current.

RESISTANCE, ELECTRICAL The resistance in a substance to the flow of electric current. Devices with a specific resistance, called resistors, are used in electronic circuits.

SEMICONDUCTOR A substance with electrical properties between a conductor and an insulator that conducts a little electricity.

SILICON CHIP A wafer-thin slice of silicon that contains thousands of microscopic electronic circuits.

SOFTWARE The information and instructions fed into a computer.

SOLAR CELL A cell that converts the energy in sunlight into electricity. Properly called photovoltaic cell.

STATIC ELECTRICITY The electricity associated with electric charges, which tends to stay where it is ("static") rather than flowing away.

TRANSFORMER An electrical device used to alter the voltage of alternating electric current.

TRANSISTOR A semiconductor device used in electronic circuits to manipulate electrical signals.

VARIATION, MAGNETIC The angular difference between magnetic north and true, or geographic, north.

VOLT The unit of electrical voltage, or "pressure," named after the electrical pioneer Alessandro Volta.

VOLTMETER An instrument for measuring the voltage.

WAVELENGTH Of a wave motion: the distance between the crest of one wave and the next. Compare to **FREQUENCY.**

XEROGRAPHY The principle on which photocopiers work. In the process, ink is attracted to electrically charged areas of a rotating drum.

Page 9
1. 300 billion 100 watt light bulbs could be powered.
2. Benjamin Franklin was a famous statesman who played a leading role in building the United States.

Page 11
1. The neutron is well named because it is electrically neutral – it has no electric charge.
2. We find sodium chloride dissolved in the sea. It is the main mineral that makes seawater salty. Table salt is sodium chloride.

Page 12
The electrically charged balloons will stick to the walls, the ceiling, and wooden or plastic doors, but not to the car. Cars are made of metal, which conducts away the electrical charges on the balloon.

Page 13
1. The energy comes from the rubbing.
2. The hairs separate because they have the same electric charge and therefore repel one another.

Investigate
Rubbing a metal spoon does not give it an electric charge, and therefore it will not pick up pins or anything else.

Page 15
The sudden expansion of the heated air along the path of lightning causes an explosion, which we hear as thunder.

Page 16
The fact that your magnet picks up a tin can does not mean that tin is magnetic. Tin cans are made from tinplate – steel sheet covered with a very thin layer of tin.

Investigate
When you tap the card holding the filings, they rearrange themselves into a pattern. This shows the lines of force of the magnetic field. In the experiment without the magnet, no pattern forms.

Page 17
The other common means of magnetic storage used in the home is magnetic tape. This is used to store music on cassette tapes and video recordings on videotape.

Page 18
You can find directions at night by looking at the stars. Two stars of the Big Dipper point to Polaris, also called the North Star because it is located nearly due north.

Page 19
1. A tricky question. In theory, the compass should point north on the Equator as elsewhere. In practice, ships are built of steel, and the huge mass of steel will affect the compass needle, making it point all over the place.
2. Displays of the Northern and Southern Lights are known as the aurora.

Page 25
A long wire and a thin wire would have the greater resistance.

Page 26
1. The two advantageous properties of gold: it can be drawn out into very fine wire without breaking, and it does not corrode.
2. The main difference between graphite and diamond is that graphite is one of the softest of minerals, while diamond is the hardest.

Page 27
Ions are atoms that have lost or gained one or more electrons. They conduct electricity as they travel between electrodes.

Page 28
Investigation
Lemon juice tastes very sour because it contains acid (citric acid). When you place the foil strips against your tongue, you feel a tingle, which is a little electric shock. The

copper and zinc have formed an electric cell, with the acid acting as an electrolyte.

Page 30

1. Fuel cells are useful on piloted space missions because they provide water for the crew to drink and wash with.
2. 240 gallons (900 liters).

Page 31

In ordinary batteries, electricity is produced as a result of chemical reaction. When the chemicals have been used up, no electricity can be produced. Batteries only have enough chemicals to last for a few hours of continuous use. Satellites need to operate for years at a time.

Page 32

Investigate

Nothing should happen when you use the distilled water, which contains no ions to conduct electricity. As soon as you add vinegar or salt, bubbles start coming from the wires, showing that electricity is passing through the solution. If you use tap water, you find that bubbles start forming right away. This is because tap water contains ions from dissolved minerals.

Page 33

Investigate

When testing the gases, if you get a "plop," the gas must be hydrogen, which burns in air. If you plunge a smoldering candle in the other gas, which must be oxygen, the candle will relight.

Page 34

Investigate

The axis of the coil in the picture points east-west. When you join the ends of the wire to a battery, you will see the compass needle move. When you carry out the iron filings experiment, you will find that you get a pattern that is similar to the one you get with a bar magnet.

Page 35

You need to use insulated wire, otherwise the current will just be conducted into the bolt rather than going around the coils. Your electromagnet remains magnetic only while the current passes through the coil.

Page 37

Light from the Sun takes 8.33 seconds to reach the Earth.

Page 39

The coils heat up because of their high electrical resistance.

Page 40

Investigate

When you push the magnet into the coil, the compass needle kicks, showing that current is flowing in the wire. The compass needle also kicks when you pull the magnet out of the coil. By moving the magnet in and out, you set the compass needle swinging.

The galvanometer is named after the pioneering electrical scientist Luigi Galvani.

Page 45

AC stands for alternating current.

Page 46

It would not be a good idea to fill the bulb with ordinary air because it contains oxygen, which would combine with the hot metal in the filament and burn out the filament.

Page 48

"Vacuum tubes" was a good name for electron tubes because they contained a vacuum – the air had been removed from them.

Page 54

Particle accelerators can't use neutrons as "ammunition" because they have no electric charge and therefore can't be accelerated by an electric field.

For further reading

Ardley, Neil.
Science Book of Electricity.
Harcourt, Brace, Jovanovich, New York. 1991.

Bailey, Mark.
Electricity.
Raintree Steck-Vaughn, Austin, TX. 1988.

Bailey, Mark.
Energy from Oil and Gas.
Steck-Vaughn, Austin TX. 1988.

Baker and Haslem Staff.
Electricity.
Macmillan, New York. 1989.

Cash, Terry.
Electricity and Magnets.
Franklin Watts, New York. 1989.

Collinson.
Renewable energy.
Steck-Vaughn, Austin, TX. 1992.

Dineen, Jacqueline.
Energy from Sun, Wind, and Tide.
Enslow Press, Hillside, NJ. 1988.

Gardner, Robert.
Energy Projects for Young Scientists.
Childrens Press, Chicago, IL. 1985.

Jennings, Terry.
Energy.
Childrens Press, Chicago, IL. 1989.

Neal, Philip.
Energy, Power Sources and Electricity.
Trafalger, North Pomfret, VT. 1989.

Index

Numbers in *italics* refer to illustrations.